AS THE

c-1

Enriquez, Juan,
As the future catches you :

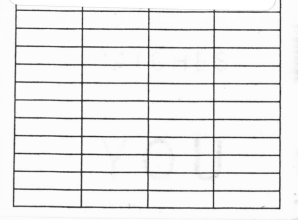

AS THE

HOW GENOMICS & OTHER FORCES ARE CHANGING

FUTURE

YOUR LIFE, WORK, HEALTH & WEALTH

CATCHES

YOU

Juan Enriquez

THREE RIVERS PRESS
NEW YORK

AS THE

HOW GENOMICS & OTHER FORCES ARE CHANGING

FUTURE

YOUR LIFE, WORK, HEALTH & WEALTH

CATCHES

YOU

Juan Enriquez

THREE RIVERS PRESS
NEW YORK

Published in the United States by Three Rivers Press, an imprint of the
Crown Publishing Group, a division of Random House, Inc., New York.
www.crownpublishing.com

Three Rivers Press and Tugboat design are registered trademarks of
Random House, Inc.

Originally published in slightly different form in hardcover in the United
States by Crown Business, an imprint of the Crown Publishing Group, a
division of Random House, Inc., New York.

Some material previously published in the Spanish language, in different
form, by Editorial Planeta Mexicana, S.A. de C.V. in 2000. Copyright
© 2000 by Juan Enriquez Cabot. Copyright © 2000 by Editorial Planeta
Mexicana, S.A. de C.V.

Library of Congress Cataloging-in-Publication Data
Enriquez, Juan.
As the future catches you : how genomics and other forces are changing
your life, work, health, and wealth / by Juan Enriquez.
1. Technology—Social aspects.   2. Gene mapping—Social aspects.
3. Civilization, Modern—21st century.   I. Title.
HM846+
303.48'3—dc21   2001028239

ISBN-13: 978-1-4000-4774-1
ISBN-10: 1-4000-4774-9

Printed in the United States of America

Design by Blue Cup Design

10 9 8 7 6 5 4 3 2

First Paperback Edition

THE FUTURE OF ANY COUNTRY DEPENDS . . .

ON BEING ABLE TO BRING UP CHILDREN LIKE DIANA AND NICOLAS . . .

FULL OF CURIOSITY, WIT, AND KNOWLEDGE . . .

ON HAVING MOTHERS LIKE MARY . . .

SMART, KIND, AND COMMITTED . . .

AND ON THEIR HAVING GRANDPARENTS WITH VALUES LIKE THOSE OF . . .

MARJORIE AND TONY, JAY AND ANNE . . .

C O N T

BEFORE
YOU
START

READING

. . .

USUALLY BOOKS ON TECHNOLOGY AND ECONOMICS ARE HARD TO READ.

INSTEAD OF WRITING A TREATISE, I WOULD LIKE YOU AND ME TO HAVE A

CONVERSATION. THERE IS SPACE ON EACH PAGE FOR YOUR OWN NOTES,

THOUGHTS, QUESTIONS, REACTIONS.

THE BOOK IS WRITTEN SO THAT ANY CURIOUS PERSON CAN READ IT. YOU DO

NOT HAVE TO BE A DOCTOR, SCIENTIST, OR COLLEGE GRAD.

READ IT ALL THE WAY THROUGH. IF THERE IS SOMETHING THAT IS HARD TO

UNDERSTAND . . . COME BACK TO IT LATER.

WHAT MATTERS ARE THE TRENDS GOING ON THROUGHOUT THE WORLD, NOT

THE SPECIFIC KNOWLEDGE AVAILABLE TODAY . . . THAT WILL CHANGE BY THE

END OF THE WEEK.

IF SOMETHING INTERESTS YOU AND YOU WANT TO KNOW MORE, I HAVE

INCLUDED ENDNOTES FOR MOST PAGES, WHICH WILL LEAD YOU TO MORE

IN-DEPTH STUFF—SOME TECHNICAL, SOME SIMPLY FUN AND INTERESTING.

HOPEFULLY, I CAN TRANSMIT SOME OF THE EXCITEMENT I FEEL ABOUT THE

MARVELOUS ADVENTURES GOING ON IN SCIENCE AND TECHNOLOGY TODAY.

AND HOPEFULLY, THIS WILL BE A FUN READ.

If you have any comments please send me a note at jenriquez@yahoo.com

AS THE

FUTURE

CATCHES

YOU

I

MIXING APPLES, ORANGES, AND
FLOPPY DISKS . . .

If it seems like your world has been topsy-turvy over the past few years . . .

Consider what's coming.

Your genetic code will be imprinted on an ID card . . .

For better and worse.

Medicines will be tailored to your genes and will help prevent specific diseases for which you may be at risk.

(But . . . your insurance company and your prospective employer may also find out that you are genetically disposed to, say, heart disease, or breast cancer, or Alzheimer's.)

> Meanwhile, lone individuals are birthing not just companies but entire industries that rapidly become bigger than the economies of most countries.
> But unlike growth industries of the past . . . cars and aerospace, for example . . . the industries that will dominate our future depend on just a few smart minds . . .
> Not a lot of manpower . . .
> So during a period of prosperity and economic growth . . .
> Wealth is ever more mobile and concentrated.

You and your children are about to face a series of unprecedented moral, ethical, economic, and financial issues.

The choices you make will impact where you live, what you earn, what your grandchildren will look like, how long you live.

It all starts because we are mixing apples, oranges, and floppy disks.

Put an orange on your desk . . .
Next to a floppy disk or CD . . .
Although each seems very different today . . .
They are becoming one and the same.

Your computer runs on a code based on "1"s and "0"s.
If you change the order and number of these 1s and 0s . . .
By tapping the keyboard . . .
You capitalize a letter, change a sentence, send an e-mail, transmit a photograph or music.
The floppy disk is simply the container for these 1s and 0s.
But it is reading and rewriting the code inside that drives change.

As of 1995, we began to read the full gene sequence of . . .
Bacteria, insects, plants, animals, humans.
It is written in a four-letter code (A, T, C, G) . . .
If you change this code, just as if you change the code in a floppy disk or on a CD . . .
You change the message, the product, the outcome.

WE ARE BEGINNING TO ACQUIRE...
DIRECT AND DELIBERATE CONTROL...
OVER THE EVOLUTION OF ALL LIFE FORMS...
ON THE PLANET....

Including ourselves.

The skin and pulp of the orange that sits on your desk . . .
Is just packaging . . .
What matters is the code contained in the seeds.
Each seed has a long string of gene data that looks like . . .

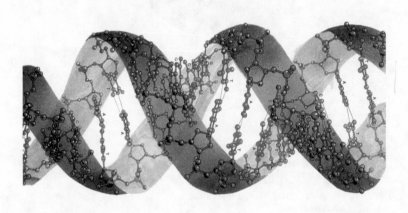

The seed guides growth, how a tree and its leaves develop . . .
The size, flavor, color, shape of fruits.
If you can read the code . . .
And rewrite it . . .
You can turn an orange into a vaccine, a contraceptive,
a polyester.
Each of these things has already been done in corn.

Today, bananas and potatoes can vaccinate you against things like
cholera, hepatitis, diarrhea.

You can harvest bulletproof fibers . . .

Grow medicines in tobacco.

And it's not just apples, oranges, and corn that are rapidly becoming
different organisms.

Mosquitoes

are flying hypodermic needles.

They can infect you with malaria, dengue, and other
awful things.

They do so by transferring a little bit of genetic code
through their saliva . . .

Into your bloodstream . . .
Which then reprograms part of the way your cells operate . . .
By changing your genetic code ever so slightly . . .
In ways that can make you very sick.
So why not engineer mosquito genes so that they have the
opposite effect?[1]

If mosquito saliva contained antibodies . . .
Or if you made it hard for malaria to mutate inside a mosquito's body . . .
You could immunize people and animals . . .
By making sure they were bitten.

Because the language of genes (A, T, C, G) is the same
for all creatures . . .
You can mix species.
If you are an artist, the genes that make jellyfish fluoresce
at night . . .
Can be used to make a bunny glow under black light.
If you are an M.D., the same genes can be placed in monkeys
to serve as markers . . .
Which help identify cures for diseases like Alzheimer's
and cancer.

By reading and rewriting the gene codes of bacteria, plants, and animals . . .

We start to turn cells, seeds, and animal embryos into the equivalent of floppy disks . . .

Data sets that can be changed and rewritten to fulfill specific tasks.

We start deliberately mixing and matching apples and oranges . . .

Species . . .

Plants and animals.

These discoveries may seem distant, abstract, more than a little scary today.

But they will change the way you think about the world . . .

Where you work . . .

What you invest in . . .

The choices your children make about life . . .

What war looks like.

Many are unprepared for . . .

The violence and *suddenness* with which . . .

New technologies change . . .

LIVES . . .
COMPANIES . . .
COUNTRIES . . .

Because they do not understand what these technologies can do.

For instance . . .

The digital revolution was a large waterfall.

In a couple of decades, the world's dominant language became . . .

Strings of ones and zeroes.

A language we cannot speak or read directly . . .

We have to use chips and machines to understand.

(This is different from communicating in, say, English-Chinese-English. No human translator can take the raw code inside a floppy disk, cell phone, pager, or TV and simultaneously tell you what it says.)

> Digital code is what drives rapid growth today.
>
> It allows mergers like AOL Time Warner . . .
>
> It drives the Internet, TV, music, finance, IT, news coverage, research, manufacturing.
>
> A few countries and companies understood this change.
>
> That is how poor countries like Finland, Singapore, and Taiwan got so wealthy . . .
>
> So quickly . . .
>
> But a lot of folks just did not learn to read and write a new language . . .
>
> And even though they produced more and more goods, particularly commodities . . .
>
> And even though they restructured companies and governments . . .
>
> Cut budgets, raised taxes, built large factories and buildings . . .

They got a lot poorer.

(In 1938 the richest country per person in Asia was . . . the Philippines. In 1954, according to the World Bank, the most promising Asian economy was . . . Burma. Both remain commodity economies . . . Both are sidelined from the digital revolution . . . And you probably would not like to live in either country.)

Your world changed when you went "On Line."

One day you used a fax or e-mail . . .

And it soon became hard to conceive of living with only snail mail.

If you understood this change early . . .

And invested or worked in some of the companies driving the digital revolution . . .

You are probably quite well off . . . (as a country and/or as an individual).

If you came late, as a speculator, without understanding what a digital language does, or does not do . . .

You probably lost a lot of money during the year 2000.

Your world . . . and your language . . . are about to change again.

The two nucleotide base pairs that code all life . . . A-T, C-G . . .

Have already led some of the world's largest companies . . .

Monsanto . . . DuPont . . . Novartis . . . IBM . . . Hoechst . . . Compaq . . . GlaxoSmithKline . . .

To declare that their future lies in life science.

They have abandoned, sold, spun off core business divisions . . .

And launched themselves into selling completely new products . . .

Which is why so many chemical, seed, cosmetic, food, pharmaceutical companies . . .

Are Partnering, Merging, Growing.

Some life-science companies will crash spectacularly . . .

Others will get larger than Microsoft or Cisco . . .

(Companies that are already larger than the economies of most of the world's countries.)

The world's mega-mergers are going to be driven by digital and genetic code.

> Consider what is about to happen to medicine.
>
> You currently spend about nine times as much for doctors and medical interventions . . .
>
> As you do on medicines and prevention.
>
> In the measure that we understand how viruses, bacteria, and our bodies are programmed . . .
>
> And how they can be reprogrammed . . .
>
> Treatment will shift from emergency interventions . . .
>
> Toward deliberate and personalized prevention . . .
>
> (Just as dentistry did.)
>
> And we may end up spending just as much on pharmaceuticals as we do on doctors.
>
> These medicines do not have to be pills or injections . . .
>
> They could be a part of the food you eat every day, your soap or cosmetics . . .
>
> Perhaps you will inhale them or simply put various patches on your skin.

(This is why Procter & Gamble is thinking of merging with a pharmaceutical company, why L'Oréal is hiring molecular biologists, and why Campbell's is selling soups designed for hospital patients with specific diseases.)

THE DOMINANT LANGUAGE . . .
AND ECONOMIC DRIVER . . . OF THIS
CENTURY . . . IS GOING TO BE
GENETICS.

Those who remain illiterate in this language . . .
Won't understand the force making the single biggest
difference in their lives.

Many countries and companies
just
don't
get
it.

They continue to invest primarily in stuff they can see and touch . . .
Even though two-thirds of the global economy . . .
Is already a knowledge economy.
They do not invest in, or attract, smart people who are science-literate.
They do not get particularly concerned as many of their brightest leave.

They forget . . .
You need ever fewer people, time, or capital . . .
To build a nation . . .
Become an economic superpower . . .
Wage war effectively . . .
Or launch a global business.
But you do need technology-literate people.

Lack of technology literacy . . .

Is one of the reasons the gap between the richest and the poorest
countries in the world is growing so quickly . . .

Why there is a 427:1 gap.

One way to get to where you want to be is to find a good map
and a smart guide.

The great cartographers of today aren't mapping continents,
rivers, mountains, or cities.

They are mapping the genetic code of all living things.

These maps are changing the way we look at all life . . .

Because they provide blueprints crucial to almost every
business.

**These maps change what we can make and
how we make it . . .**

(One sad consequence of science moving so quickly in so many different
areas is that you find ever fewer examples of good science fiction
today . . . The future simply catches up to the imagination too quickly . . . As
of January 31, 2001, it was legal to clone human embryos in Britain.)[2]

Willingly, or unwillingly, the ability to read the human gene code
(genome) . . .

Makes us all explorers.

We have to beware of waterfalls.

Some will have a sense of how to navigate . . .

Others will drift along placidly . . .

Till the water gets rough and murky . . .

And the bottom suddenly disappears.

Those who approach waterfalls in canoes . . .

With no map . . .

Are unlikely to survive if the waterfall is really high.

This happens to families, regions, and whole countries . . .

Particularly when they fail to make the next generation smarter . . .

Or decide to exhaust themselves by desperately paddling against change . . .

They too become irrelevant and disappear.

YOUR FUTURE,

THAT OF YOUR CHILDREN,

AND THAT OF YOUR COUNTRY DEPEND ON...

UNDERSTANDING A GLOBAL ECONOMY DRIVEN BY TECHNOLOGY.

UNDERSTANDING CODE, PARTICULARLY GENETIC CODE, IS TODAY'S MOST

POWERFUL TECHNOLOGY.

This book is a rough map . . .

To help you understand a new language . . .

A new code . . .

Because on February 12, 2001, anyone with access to the Internet . . .
Could suddenly look at a new atlas . . .
One containing the whole human genome.

[
(We all know what

October 12, 1492,

means . . . but Columbus never knew, and we still argue today over just where he landed . . . It took decades to realize how a new world map would end up changing the balance of power among all countries . . . Often we do not realize the power of a newly minted map . . . But grammar schools teaching future generations will recall February 12, 2001, even though we are just starting to explore.)[3]
]

II

THE 427:1 GAP

David Walton, a third-year Harvard medical student . . .
Went to Haiti for a clinical internship . . .
Was shocked by the poverty and disease . . .
Wrote a long, emotional e-mail to a lifelong friend . . .
And got a one-line response . . .

"Sounds pretty depressing. What else is going on down there?"[1]

Extreme income gaps . . .
Are so common . . .
And so large . . .
That we rarely question . . .
Why they occur . . .
And why they are increasing so rapidly.

An African folk tale . . .

Every morning, a gazelle wakes and thinks, "To stay alive, I have to run faster than the fastest lion."

There is always a significant incentive for gazelles (and small or new countries) to learn to run faster.

Just over the hill . . .

A lion realizes, "I have to run faster than the slowest gazelle, or I'll go hungry."

Even lions (if they get lazy or if gazelles get smarter) can starve . . .

(So can great empires.)

Many countries, regions, and people learn the consequences of this folk tale the hard way.

Three-quarters of the . . .

FLAGS, BORDERS, ANTHEMS, AND MONEYS represented at the United Nations today . . .

Did not exist fifty years ago.

States are falling apart at an unprecedented rate . . .

Because governments and citizens do not understand . . .

Why technology is relevant to their daily lives . . . and how it changes their future.

Even within the United States . . .
During a period of unprecedented growth . . .
And technological leadership . . .
Middle-class bankruptcies . . .
Increased from 313,000 in 1980 to 1,661,996 in 2003.
Two-thirds of these were people who lost their jobs . . .
And found it impossible to catch up . . .
With a rapidly changing economy.[2]

(According to Visa International, bankruptcies increased thirty percent in the first four weeks of 2001 when compared with the same period a year ago . . .)

As a country or region...
If you understand...
And apply technology...
You can postpone...
Becoming a wondrous...
But abandoned...
Archaeological site.

ANGKOR WAT

This is a relevant skill . . .
Those who forget . . . are often forgotten.
After all . . .
The cradle of civilization is today's Iraq.
In 1200, Cambodia was one of the richest countries in the world . . .
(which is why Angkor Wat is so spectacular).
In 1500, Peru and Mexico awed Europeans . . .
As late as the 1960s, the ''Switzerland of the Middle East'' . . . was Lebanon . . .
And the ''Switzerland of Africa'' . . . was Uganda.

When a country or region is consumed by internal political battles . . .
And ignores the technologies that sweep through other states . . .
It becomes irrelevant.

(Although it can still inspire wonderful poetry):

Ozymandias

"I MET A TRAVELLER FROM AN ANTIQUE LAND
WHO SAID: TWO VAST AND TRUNKLESS LEGS OF
STONE STAND IN THE DESERT. NEAR THEM, ON
THE SAND, HALF SUNK, A SHATTERED VISAGE
LIES, WHOSE FROWN, AND WRINKLED LIP, AND
SNEER OF COLD COMMAND, TELL THAT ITS SCULP-
TOR WELL THOSE PASSIONS READ WHICH YET
SURVIVE, STAMPED ON THESE LIFELESS THINGS,
THE HAND THAT MOCKED THEM, AND THE HEART
THAT FED; AND ON THE PEDESTAL THESE WORDS
APPEAR: "MY NAME IS OZYMANDIAS, KING OF
KINGS: LOOK UPON MY WORKS, YE MIGHTY, AND
DESPAIR!" NOTHING BESIDE REMAINS. ROUND
THE DECAY OF THAT COLOSSAL WRECK, BOUND-
LESS AND BARE THE LONE AND LEVEL SANDS
STRETCH FAR AWAY."

SHELLEY

(1817)

(No wonder they say the last words of a civilization are
its buildings and monuments... Percy Bysshe Shelley
wrote "Ozymandias" inspired by the egomaniacal
Pharaoh Ramses II. Make sure you give a copy of the
poem to your local tyrant or blowhard.)

The consequences of ignoring technology . . .

Or thinking things can just go on . . .

Are more severe . . .

Than ever before.

You do not see these much anymore:[3]

FIRST *Official Flag of the Confederacy* (1861)

LAST *Flag of the Soviet Union* (R.I.P. 1991)

(Just two out of hundreds of symbols and countries that survive only in museums and history books.)

But why is it so hard to keep a country together today?

Individuals can leverage technology. They no longer need a big state to get very rich.

Because the way you generate wealth is very different today.

In 1750, someone working in the world's richest country was about five times wealthier than someone working in the poorest one. As long as economic development depended primarily on agriculture, it was hard for one region's work to be far more valuable than that of its neighbor.

Perhaps a more disciplined population—one that got up to milk cows earlier and had better land and better leaders—could accumulate a little more wealth.

But not a great deal more.

(Through the eighteenth century, Europe's average yearly economic growth was about 0.07 percent per year.)[4]

ALL THIS CHANGED WITH THE
INDUSTRIAL
REVOLUTION.

Suddenly one person's labor could be multiplied a hundred- or a thousandfold, because it was
a machine doing most of the work, not one individual's hands.

Countries that developed machines became far richer.
A civilization with a great history, culture, and pedigree, but that did not find a way to multiply its citizens' output, became far poorer.

In 1840, just as the Industrial Revolution was beginning,
two great states, China and India, accounted for 40 percent of world trade.

These two countries continued producing the best and most luxurious handmade goods in the world . . . Silks, jewels, jade.

Meanwhile, Europe and the United States began producing far more products. And each product was getting cheaper.

When Henry **Ford** built his first Model T in 1908, he sold it for $900 . . .

There were cars that were more luxurious, better made, or cheaper . . .

But Ford industrialized and standardized mass production . . .

("The customer can have any color he wants, so long as it's black," he said.)

Four years after starting production, a Model T cost $690, one-quarter less than when it was launched.

Two years later, just before World War I, the car sold for $490 . . .

And Ford was selling more than one million cars per year.

By 1925, Model Ts were $290 . . .[5]

But neither **India** nor **China** was industrializing.

Today the average Indian lives in extreme poverty . . .

Although surrounded by beautiful palaces and brocades.

China and India together publish 3.9 percent of the world's science papers . . .

AND CHINA AND INDIA . . .
TOGETHER . . .
NO LONGER REPRESENT . . .
40 PERCENT OF WORLD TRADE . . .

JUST 5.5 PERCENT.[6]

(This is why, even in 1998, the average South Asian income per person was lower than that of sub-Saharan Africa, $430 per year vs. $510.)

Perhaps one can understand why the Chinese were somewhat arrogant and blind to change.
Of the fourteen dynasties, ten lasted longer than the entire history of the United States . . .
They developed paper, the compass, fine porcelain, gunpowder, movable type, guns.

Yet they frittered away every technological advantage . . .
Because they did not trust their own people . . .
And they feared new ways of doing things.
China's first emperor, Qinshihuang, set the tone by burning all books . . .
To erase what had come before.

Something similar occurred in the Americas.
In 1800, Cubans and Argentines were richer than Americans.
But the United States educated its population . . .

> **Built infrastructure . . .**
>> **Accumulated capital . . .**
>>> **And adopted high-tech methods . . .**
>>>> **In agriculture and textiles . . .**
>>>>> **Many benefited.**

Despite a civil war that killed more than 600,000 souls, the United States started becoming a global agricultural power.

In 1850, it began using grain elevators, which . . . COST ONE-TENTH AS MUCH AS USING PEOPLE TO LOAD AND UNLOAD . . .

By 1890, mechanized farming had become . . . ONE-FIFTH AS EXPENSIVE AS USING HORSES.[7]

Meanwhile in Mexico, one of the first countries in the hemisphere with a university or a printing press, few people were aware of or participated in the Industrial Revolution. Even toward the end of the century, when dictator Porfirio Díaz decreed science and technology priorities, few Mexicans understood or were given the instruments necessary to change the way they made their living.

THROUGHOUT MANY REGIONS OF LATIN AMERICA...

EDUCATION... THE INDUSTRIAL REVOLUTION...

CAME LATE... IF AT ALL.

Japan lived through something similar.

The island isolated itself from the world, and focused on controlling its own people.

There was little incentive to study . . .

Or learn more . . .

Than that appropriate to one's status.

Or to adopt a technology that would change one's status.

> Until Commodore Matthew Perry showed up with his black ships in 1853 . . . and forced the Japanese to confront the world.

> Even then, it took over a century for Japan to catch up.

Through 1935, the difference between what someone in Japan produced and what a Mexican or a Brazilian produced was minimal.

IN 1960, "MADE IN JAPAN" WAS A SYNONYM FOR BAD QUALITY.

> But, unlike much of Latin America, Japan eventually understood the value of science, technology, and educating one's own. Today, Japanese children do far better on standardized science tests than do Americans.

(In the 1990s, Japanese and European science literacy greatly exceeded that of U.S. children after fourth grade.)[8]

> And despite a world war that destroyed much of Japan at midcentury . . .
>
> **A citizen of Japan produces five times as much wealth today as does a Latin American.**

Science and technology allow people to multiply their productivity much faster than those who do not have the same knowledge or instruments.

And now the difference between what one person produces in the richest and the poorest countries in the world is no longer **5** to **1** . . .

It is **4271** to **1.**

(And with the IT and genetics revolutions . . . the difference will soon be more than **1,000** to **1.**)

(As you leverage technology, you produce more wealth faster. Try looking at this in terms of a Big Mac. The average American works seven minutes to buy one of these hamburgers, a Canadian works 8, Spaniards 11, Mexicans 60, Indians 137, and Brazilians 215 minutes. In 1940 it took a Californian the equivalent of twenty-seven minutes to buy a McDonald brother's one-eighth-pound burger.)[9]

III

THE NEW RICH...AND THE NEW POOR

DIFFERENCES IN EDUCATION, PARTICULARLY IN SCIENTIFIC LITERACY, CAN LEAD
TO EXTREME CHANGES IN RELATIVE WEALTH AND POSITION...V
E
R
Y...QUICKLY

Fifty

years ago, a Taiwanese was governed by the remnants of a corrupt and drug-riddled government that had lost China . . .

And retreated to the country's poorest and most isolated province.

Those in *Mexico produced, on average, twice that of someone living in Taiwan.*

By 1974, Taiwan had imposed brutal university entrance examinations and emphasized scientific literacy. The industrial plant grew, as did exports and competitiveness . . .

And the average *Taiwanese produced twice as much as a Mexican.*

By the end of the 1990s, when one visited Taiwan, one could walk through street markets similar to those of Latin America, and see a myriad of small booths with signs like

"Buy Here!!"

"Sale,"

"Much Cheaper."

But these markets were not selling fruits, vegetables, or piñatas . . .

They were selling the components that are used to assemble computers.

Taiwan had become one of the world leaders in PC and chip manufacturing . . .

And a *Taiwanese was producing four times more wealth than a Mexican.*

These differences are going to increase...

Not just between Taiwan and Mexico . . .

But between countries . . .

And within countries . . .

Because there are regions whose economy and people . . .

Are still based on farming . . .

Or assembling basic products . . .

And there are other regions . . .

Where people are generating vast amounts . . .

Of new knowledge.

Factory workers who upgrade skills and technology rapidly increase income.

DIVERGING HOURLY WAGES[1]

(total dollar wage for manufacturing workers in U.S. $)

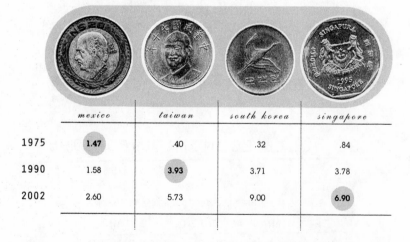

	mexico	taiwan	south korea	singapore
1975	1.47	.40	.32	.84
1990	1.58	3.93	3.71	3.78
2002	2.60	5.73	9.00	6.90

(Meanwhile, back in the United States . . . a factory worker made the equivalent of $0.15 per hour in 1897, by 1975 it was $6.36, and in 2002 it was around $21.11. It is not that Americans are working harder. The average person worked 52 hours in 1900 versus 37.9 in 2000. The difference is that one educated individual can produce a lot more.)[2]

The world's economy has changed.

It used to be primarily *agricultural.*

Then *manufacturing* grew quickly . . .

By the 1960s, agriculture, manufacturing, and services each accounted for about a third of the global economy.

Now a vague area . . . mislabeled *services* . . . is becoming dominant.[3]

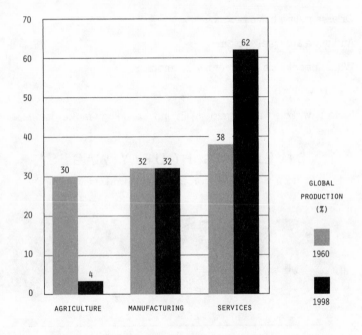

THE WORLD ECONOMY 1960–1998

AFTER INVENTING THE PLOW, WE REMAINED FARMERS FOR 5,000 YEARS...

THEN WE RAN FACTORIES FOR 150 YEARS...

NOW WE ARE IN THE IDEA BUSINESS.

WE MAKE MONEY PRIMARILY BY

MANIPULATING NEURONS.[4]

Many believed . . .

And many still believe . . .

That those working in services cannot make much money . . .

Because they don't "build things that are real."

Some working in the service economy are flipping hamburgers . . .

Or guiding tourists . . .

But most value created within the service economy . . .

Is knowledge-based.

And those who generate knowledge are the ones getting richer.

As a product becomes standardized and is mass-produced, be it a seed, a machine, or a computer program . . .

The **knowledge component becomes more important** . . .

And manual labor less valuable.

Even in large and complex manufacturing processes like building automobiles . . .

Transport and insurance costs exceed those of steel . . .

Or of manual labor.[5]

(A Ford Taurus has more than 120 computer chips . . . more computing power than the *Apollo* lunar excursion modules.)

The same is true even in agriculture . . .

Look at something as basic as flower production.

Imagine they asked you to find the ideal country for exporting flowers.

You might be tempted to go looking for a place that had . . .

> A lot of land . . .
> Cheap labor . . .
> Fertile soil . . .
> Warmth . . .
> Sun . . .

You would be wrong.

Brazil is not the world's greatest flower exporter . . .

It is Holland . . .

A small, fog-bound country, part of whose land is lower than the cold North Sea.

Because even in something as basic as agriculture, what matters is knowledge-based services.

Schiphol is Europe's most efficient airport (and it is expensive to keep flowers waiting).

Dutch universities honor horticulture professors . . .

Banks lend with tulip bulbs as guarantees . . .

Cities of greenhouses create their own weather . . .

Holland's markets trade and deliver a standardized product and quality globally.[6]

In the old economy . . .

Building a billion-dollar fortune required . . .

Decades of hard work . . .

A powerful host country. . . .

Thousands of workers . . .

And thousands of storefronts.

Today, a kid with a smart idea, a couple of friends, and some luck can make a lot of money . . .

Very quickly.

(In 1999, an 18-year-old college dropout started fiddling with his laptop and came up with a program that would allow you to download your favorite music from the Internet . . . Thirty-eight million kids thought this was a great idea . . . Record-company execs had a heart attack . . . College computer centers crashed because 40 percent of their bandwidth was used to download music . . . Lawsuits flourished . . . But in the words of a music industry CEO, thirty-eight million people "can't all be criminals." . . . So one of the world's largest media conglomerates ended up partnering with Napster . . . and now there are a myriad of legal and gray-area file-sharing programs.)

The rules of an economy

Based on knowledge and networks

Are very different from those

Of a manufacturing-based economy.[7]

In the old economy . . . if something was *scarce* . . . it was *valuable.*

Those who controlled the mines, owned the exclusive rights to a product, or had the only copy of something could become very wealthy.

Today it is just the opposite.

(Of course, there are still some fuddy-duddy economists using ever more complex equations to try to convince you that some basic economic "laws" haven't changed . . . that there is no new economy.)

Even though there are still many exclusive gewgaws
we can buy . . . With a brand name . . . Expensive . . .
These goods are not what drive most economic growth.

And therefore those who control these products . . .
are no longer the richest people in the world.

WHEN YOU ARE TRYING TO SPREAD, AND SELL,
KNOWLEDGE . . . KEEPING SOMETHING
"EXCLUSIVE" AND "RARE" OFTEN
LEADS TO A LOSS OF VALUE. WHAT
MATTERS MOST IS THAT THE PURCHASER
BECOMES PART OF A NETWORK . . . AND THAT
THE NETWORK KEEPS GROWING.[8]

The first purchaser of a telephone, or a fax machine,
had a useless product . . .

He could not communicate with anyone.

Each additional phone or fax sold makes the network
more valuable . . .

So each purchaser becomes a salesman . . .

"Don't you have a fax? You should get one."

Now you do not need a business, government, or fortune to set up a very large network . . .[9]

Just word of mouth.

Look, for example, at one of the world's largest computer networks.

Outer space is a very noisy place . . .

Radiation from various exotic objects like stars, black holes, quasars, and supernovas permeates almost everything.

So those interested in finding out whether there are any orderly transmissions within this background din . . .

And therefore discovering other intelligent life forms . . .

Must compare and analyze trillions of data points . . .

Something no single computer on Earth could do.

So they ask PC owners like you whether you will donate part of your unused capacity . . .

Because even if you type two letters per second, you are using only a fraction of your PC's processing power.

And the unused parts of a computer's brain can parallel-process signals from space . . .

If you are logged on to the Internet.

By now you may be thinking, what a wacky idea—who on Earth would have supported a project like that?

Well, among others . . .

NASA Ames Research Center
NASA HEADQUARTERS ✻ The National Science
Foundation ✻ THE U.S. DEPARTMENT OF
ENERGY ✻ The U.S. Geological Survey ✻ NASA
JET PROPULSION LABORATORY
The International Astronomical Union
ARGONNE NATIONAL LABORATORY
The Alfred P. Sloan Foundation ✻ THE DAVID
AND LUCILE PACKARD FOUNDATION
The Paul G. Allen Foundation ✻ THE MOORE
FOUNDATION ✻ The Universities Space
Research Association ✻ THE PACIFIC SCIENCE
CENTER ✻ The Foundation for Microbiology
SUN MICROSYSTEMS ✻ Hewlett-Packard.

Those who founded Search for Extraterrestrial Intelligence (SETI) hoped 100,000 global citizens would sign up . . . More than two million did. And *SETI@home* is now the world's largest community computer project . . . With 280,000 years of processing time available . . . And a network that grows every week.[10] AIDS researchers soon copied this model, linking more than 100,000 PCs in over eighty countries . . . Creating the world's largest academic computing site . . . Designed to understand and attack a virus that mutates very rapidly.[11]

(And of course, volunteer networks can morph into business networks. Distributed Science and Popular Power are farming out large computational jobs to individuals and paying for the use of their computers.)

The broader the network . . .
The easier it is to communicate . . .
The more each product is worth . . .
And the cheaper it becomes to purchase or use it.

And it is getting much cheaper to transmit and acquire information because . . .

EVERY EIGHTEEN MONTHS . . .
THE COST OF A COMPUTER CHIP DROPS . . .
AND ITS ABILITY TO PROCESS DATA . . .

DOUBLES.

(This concept, referred to as Moore's Law, is based on an observation on the miniaturization of transistors made by the founder of Intel, Gordon Moore. Today's $900 laptops are thirteen times as powerful as IBM's 1970 main-frame behemoths, which cost $4,700,000. Over the past few years, computers have gotten so powerful that many questioned whether there would not be a quantum limitation to continued miniaturization, but in January 2000 IBM scientists showed that you could theoretically build a computer on a quantum scale.)

This computer (1944): Cost 200 times as much as the computer that sits on your desk . . . in constant dollars:

And your small computer, if bought in the last two years, is at least 100,000 times more powerful.

It is not just computers that are getting cheaper.[12]

MORE . . . CHEAPER . . . FASTER
(cost in constant U.S. dollars)

DEVICE	COST AT INTRODUCTION		COST TODAY
Handheld calculator	120	1972	5
Color TV	1,000	1954	180
VCR	1,395	1978	80
Cell phone	4,195	1984	Free (with activation)

(The decrease in costs and increase in relative wages become even more apparent if you put them in terms of how long you have to work to buy a TV: in 1954 around 562 hours . . . 1971 about 174 . . . 1997 less than 24 hours.)

As the machine, or the chip, that carries out various computations becomes less and less expensive, it is simultaneously becoming more valuable.

(Which is why 0 percent of U.S. households were connected to the Internet in 1990 . . . In 2003 over 53 percent were networked . . . and people were spending more than one hundred billion minutes on line . . . every month.)

What makes money today is the knowledge necessary to process and distribute information.

Look at software, for instance.

The first copy of a major computer program costs millions . . .
The second costs pennies.

The more people who use the program . . .

The easier it is to exchange information . . .

The more valuable the program becomes . . .

And the richer the company and its employees.

Microsoft was founded in 1975.
And generated hundreds of employee millionaires . . .
and various sundry billionaires.

(I dare you to try buying waterfront property in Seattle . . . Never mind the $30.8 million Leonardo da Vinci notebook that Bill Gates bought in 1994.)

But before Microsoft became the behemoth it is today . . .

Apple built a simpler and better operating system . . .

But it did not share . . . *It kept its program "exclusive."*

Programmers found it easier to work with Microsoft's "open" system . . .

So today you can buy 70,000 Microsoft-compatible programs . . .
and 12,000 Apple programs . . .

Even though it had a better product . . . Apple lost.

And # Microsoft set the industry standard
for the next generations . . .

Through a new language . . . First DOS . . . Now Windows . . .

Began charging a little more every time the code was modified . . .

Became the world's most valuable corporation . . .

And bought up a chunk of Apple . . .

For dessert.

Bill Gates retired as CEO at age 44. Tired of running the company day to day . . . He remained chairman of the company and tried to focus on the next stages of the information revolution . . . how to stay ahead of Oracle, Netscape, AOL Time Warner.

When Gates "retired," his personal fortune exceeded the total U.S. gold reserves held in Fort Knox, twice the number of dollar bills in circulation.

Or look at it another way . . .

In 1999 Gates' personal wealth exceeded the total value of everything produced by all the people living in Israel, Malaysia, or Chile over the course of a year . . .

Or that produced in any one of 141 of this planet's countries.

(Microsoft did its job so thoroughly that the U.S. Justice Department attempted to break up what it saw as a global monopoly. By the end of 2000, Microsoft's share price fell from $120 to the mid-$40s, which meant the company was worth a mere $246 billion.)

In a tech-driven economy . . .
The speed with which fortunes appear and disappear . . .
And waterfalls of data behind these new fortunes . . .
Is staggering.

On his thirty-fifth birthday, Jeff Bezos was worth close to $10,000,000,000.

Five years earlier, he lived in a 500-square-foot apartment.

He is primarily a bookseller . . .

But he understood the power of a new medium and a new technology, the Internet, to change commerce.

Which is why Amazon.com was touted as one of the world's most successful companies . . .

For a while.

But technology moves so quickly that either his company gets a lot bigger . . .

And a lot smarter . . .[13]

Over the next two years . . .

Or it is likely that Bezos will no longer make the list of the 100 richest Americans . . .

Much less be named, as he was in 1999, *Time's* "Man of the Year."

(During 2000, Amazon lost over 85 percent of its market value . . . oops.)

But regardless of how one individual or one company fares . . .
The overall concept is clear . . .

[KNOWLEDGE,

NOT TO MENTION THE ABILITY TO USE KNOWLEDGE,

MATTERS.]

In January 2000, before its close encounter of the judicial kind, Microsoft
was valued at around $592,000,000,000 . . .[14]
About ten times what Brazil exported in 1998 . . .
Or five times more than the total exports of the United States' second-
largest trading partner, Mexico.

But Brazil and Mexico had 171,853,126
and 100,294,036 people...
Microsoft employed 32,902 people.

A decade out...
Guess who could get a lot richer...
(And who will get a lot poorer).

Tradition used to matter . . .
Old money lasted . . .
No more.

The richest people in the United States today are very different from those of the 1980s.
Of the ten richest and their families in 1986 . . .
Only four (Walton, Buffett, Mars, and Pritzker) . . .
Made the top ten a generation later.

GREAT WEALTH 1986 VERSUS 2004[15]

(estimated wealth in billions in parentheses)

RICHEST IN U.S. 1986	MAIN SOURCE OF WEALTH	RICHEST IN U.S. 2004	MAIN SOURCE OF WEALTH
Sam Walton (4.5)	Retail	Walton family (100)	Retail
Mars family (4.0)	Candy	Bill Gates (47)	Software
John Kluge (2.5)	Media	Warren Buffett (43)	Investments
Ross Perot (2.5)	Information	Mars family (32)	Candy
Pritzker family (2.3)	Finance	Cox family (22)	Media
David Packard (2.0)	Computers	Paul Allen (21)	Software
Warren Buffett (1.4)	Value investing	Larry Ellison (19)	Software
Leslie Wexner (1.4)	Retail	Ned Johnson (15)	Finance
Lester Crown (1.3)	Inheritance, industry	Pritzker family (15)	Finance
Gordon Getty (1.2)	Inheritance, oil	Michael Dell (13)	Computers

WHOLE COUNTRIES,
REGIONS, INDUSTRIES, AND INDIVIDUALS
ARE BECOMING RELEVANT OR IRRELEVANT
IN A FEW DECADES OR YEARS.

In 1900, ten of the twelve largest U.S. corporations sold commodities.

In 2004, only three were close to selling commodities. The other nine were in manufacturing, finance, and high tech.

THE LARGEST U.S. CORPORATIONS[16]

U.S. IN 1900	U.S. IN 2004	U.S. NEXT DECADES (INDUSTRIAL SECTORS)
American Cotton Oil	Wal-Mart	
American Steel	Exxon Mobil	Software
American Sugar Refining	General Motors	Aerospace
Continental Tobacco	Chevron-Texaco	Genomics
Federal Steel	Conoco-Phillips	Bioinformatics
General Electric	Citigroup	Nanotechnology
National Lead	IBM	Photonics
Pacific Mail	AIG	Micro Materials
People's Gas	Hewlett-Packard	Finance
Tennessee Coal & Iron	Verizon	IT
U.S. Leather		Robotics
U.S. Rubber		Entertainment

But the 2004 rankings are by sales.

In general, the companies that grow fastest are those selling ideas, not those with the most assets.

General Motors was incorporated in 1916 . . .

Became the largest U.S. company . . .

And sells in fifty-two countries.

Microsoft, founded in 1975, operates in seventy-nine . . .

And rising.

How wealth is created within the United States and some Asian countries is quite different from what still happens in much of the rest of the world.

Most European and Latin American fortunes are not based on high tech...

They are inherited...

(Although a few heirs have grown the companies significantly.)

Seven out of the ten richest Europeans are what *Forbes* calls "coupon clippers"...

(That is, they are not involved on a daily basis with building up their companies.)

RICHEST EUROPEANS AND LATIN AMERICANS[17]

(estimated wealth in billions of dollars)

EUROPEAN BILLIONAIRES	WEALTH COMES MAINLY FROM	LATIN AMERICAN BILLIONAIRES	WEALTH COMES MAINLY FROM
Theo and Karl Albrecht (41)	Supermarkets	Carol Slim (14)	Telephone privatization
Liliane Bettencourt (19)	Inheritance, cosmetics, food	José and Moisés Safra (4.7)	Inheritance, finance
Ingvar Kamprad (19)	Retail	Gustavio Cisneros and family (4.6)	Inheritance, media
Rausing family (17)	Inheritance, packaging	Lorenzo Mendoza and family (4.1)	Inheritance, drinks
Quandt family (19)	Inheritance, automobiles	Gerónimo Arango (4)	Retail
Bernard Arnault (12)	Retail	Andrónico Luksic and family (3.4)	Mining
Silvio Berlusconi (10)	Media	Lorenzo Zambrano and family (3.1)	Inheritance, cement
Amancio Ortega (9)	Textiles	Eliodoro Matte and family (2.8)	Inheritance, mining, wood
Gerald Cavendish Grosvenor (9)	Real estate	Anacleto Angelini (2.5)	Energy
Stefan Persson (9)	Inheritance, retail	Eugenio Garza Lagüera and family (2.5)	Inheritance, drinks
Rudolf August Oetker and family (7.5)	Food	Alberto Bailleres (2.3)	Inheritance, mining, wood

The United States' ability to attract the world's best brains and the shift toward a knowledge economy create rapid shifts.

In 1990, not one of the world's ten wealthiest individuals was American . . .

In 2004, eight out of ten were American.

MUCH OF THE WORLD'S NEW WEALTH IS CREATED BY KNOWLEDGE . . . BUT MOST OF THE WORLD'S POPULATION STILL WORKS IN BUSINESSES OR ENDEAVORS THAT PRODUCE, ASSEMBLE, OR SELL COMMODITIES . . . SO THE GAP BETWEEN THOSE WHO ARE TECHNOLOGY-LITERATE AND THOSE WHO ARE NOT COULD EASILY **WIDEN** AS RESEARCH AND DEVELOPMENT ACCELERATES.

Many individuals, ethnic groups, and countries can quickly get left behind.

Most large Latin American corporations still sell primarily commodities.
Toward the end of the twentieth century, only three out of the top fifty were primarily high tech.[18]

(Many governments have yet to understand the logic of a knowledge-driven economy . . . They still do not realize that in the age of information, hard work, by itself, is not sufficient.)[19]

Never mind Africa . . .

(According to **Wired,** there are fewer phone lines on the African continent than there are in Manhattan . . . and half of the world's population has never made a phone call. Out of sight, out of mind?)

It is getting harder to maintain the value of the currency in these regions, because what they produce is less valuable . . . And this has consequences.

Through the mid-twentieth century . . .

The great powers of the world . . .

>>> **Still worried about . . .**

>>> **And fought over . . .**

>>> **Africa . . .**

>>> **Its people . . .**

>>> **Its territory . . .**

>>> **Its resources.**

>>> **Today ever fewer care what happens to the continent . . .**

>>> **Many countries have ceased to exist de facto . . .**

>>> **Whole generations are dying of AIDS . . .**

>>> **Genocide is common.**

And . . .

Since Africa has become irrelevant to the knowledge economy . . .

Most people have abandoned an entire continent to its fate.[20]

(From 1996–98, three out of four Somalians were underfed, 68 percent of those in Burundi, 61 percent of those in the Congo . . . And neither starvation nor AIDS may end up being the biggest problem; by 2025 smoking could kill more Africans than AIDS, TB, malaria, car crashes, and homicides do together.)[21]

The countries that get rich . . .

Are the ones attracting the great minds . . .

Or those educating their own.

Most of Africa is not . . .

And as long as it does not . . .

It has little hope . . .

Of accomplishing very much beyond surviving . . .

If that.

Let's hope the same does not occur within the Americas . . .

But it is something that should concern us, because . . .

MANY
COUNTRIES
ARE FAILING,
FALLING APART,

AND DISAPPEARING.

IV

EMPIRES OF THE MIND

[
"THE EMPIRES OF
THE FUTURE ARE
THE EMPIRES OF
THE MIND."
]

WINSTON
CHURCHILL

Rich countries no longer need great deposits of gold or diamonds . . .

Or an abundance of land . . .

Or millions of people . . .

THEY NEED TO EDUCATE THEIR POPULATION.

NO MORE . . .

THEY NEED SMART AND ENTREPRENEURIAL PEOPLE.

NO LESS . . .

THEY NEED A GOVERNMENT THAT PROVIDES
ECONOMIC AND POLITICAL STABILITY.[1]

In past centuries . . .

When economic development was dependent on agricultural
production . . .

Or on massive industrialization . . .

Having a lot of natural resources and a lot of land . . .

Was an advantage . . .

Today it is a disadvantage.

During the twentieth century, developing countries with a lot of natural resources . . .

Have been economic and administrative disasters.

The now-extinct **Soviet Union** was, until recently, the largest and richest country in the world, in terms of natural resources.

No border contained more gold, oil, uranium, or forest.

(The distance from New York to Moscow is less than from Moscow to parts of the east coast of Russia.)

The Soviet government decreed that the empire's power depended . . .

On exploiting natural resources.

People became an instrument . . .

To be used and controlled . . .

Not the main means to generate wealth.

The great scientists, engineers, and mathematicians worked on secret projects . . .

And found it hard to travel, communicate, and apply what they knew.

To photocopy an article . . . or to start a new business . . .

was **illegal**.

In a world that increasingly valued *Entrepreneurship* . . .

Communication . . .

Freedom . . .

The **Soviet Union** got poorer and poorer . . .

Until it went broke . . . and b r o k e a p a r t.

Today . . .

THE FLAG . . . BORDER . . . AND

ANTHEM . . . OF THE SOVIET STATE . . . NO LONGER SURVIVE.

(Sergei Bubka, an extraordinary pole vaulter, won the world championships three consecutive times. Despite having lived in the same town his whole life, each time he stood on the podium to receive his medal he ended up saluting a different flag and hearing a different anthem . . . And, through 2000, champion Russian athletes stood mute because their new government had yet to decide what words went with its national anthem.)

The Soviet Union is no exception . . .

Rather, it is one of many examples . . .

Nigeria has oil . . .

Indonesia precious woods . . .

South Africa diamonds and gold . . .

Brazil jungles and minerals . . .

Argentina vast fertile lands . . .

Congo minerals and gems . . .

Mexico silver and oil . . .

Colombia emeralds . . .

Saudi Arabia oil . . .

Venezuela oil . . .

Each is a great country . . .

And most of the inhabitants of these great countries are poorer today than they were twenty years ago.

An autopsy of these "rich" states is sometimes carried out within the manicured lawns of the Harvard Business School . . .

> Where a soft-hearted curmudgeon rants . . .

> And takes on assorted global bureaucrats, economists, and colleagues.

> You have to listen closely to realize that Bruce Scott . . .

> Is one of the Democrats on campus . . .

> And that he spends his life trying to help people . . .

> With the same tough love he showed when he co-edited . . .

> *U.S. Competitiveness and the World Economy* . . .

> Which was a catalyst for the massive restructuring of U.S. businesses in the 1980s.

(His partner in crime was a wonderful patrician Republican with a heart of gold, George Lodge . . . who has gone from journalism and politics to one of the distinguished chairs at Harvard. Bruce and George trained a lot of those responsible for the Asian miracle.)

Bruce teaches those focused on making a buck or two . . .

That you do not need a lot of resources to do well . . .

That many countries . . .

Have only people.

Many are small . . .

Some are not even self-sufficient in water . . .

Much less food, minerals, and fuel.

Some even lack a common history and culture . . .

And these countries had only two options . . .

Remain poor . . .

Or educate their populations.

Those that were able to follow the second path . . .

WERE SOMETIMES ABLE TO CREATE A KNOWLEDGE ECONOMY AND ARE NOW INHABITED BY SOME OF THE RICHEST PEOPLE ON THE PLANET.

In 1950, Singapore was an isolated, tiny, poor island . . . It wasn't even a country before 1965.

Its future was so bleak that its leaders went to Malaysia and asked whether it would be willing to absorb Singapore and make it part of its state.

Malaysia's leaders decided that absorbing Singapore would make their country poorer and declined the offer.

Ghana, Burma, and Sri Lanka seemed richer, far more promising countries than Singapore.

And in August 1965, *The Sydney Morning Herald* concluded: "An independent Singapore was not regarded as viable three years ago. Nothing in the current situation suggests that it is more viable today."

Singapore had no choice but to educate its people, reform its government, attract knowledge, and get to work.

Which it did . . . Led by an extraordinary man, Lee Kuan Yew. He started out as a legal adviser-negotiator on behalf of Communist labor unions . . . Gave that up in favor of technology and capitalism . . . Focused on education and talent . . . Made English the working language in 1978 . . . Attracted foreign investment.

By 1975, General Electric was the country's single largest employer, and Singapore was recognized as a leader in electronics manufacturing.

By 1985, Singaporeans were producing $8,116 per person . . . Their ex-colonial masters, the British, were at $11,237.

By 1999, Singaporeans were 2 percent wealthier than the Brits.

(Not bad for a country that did not have a national anthem before the late 1960s . . . and where a single generation grew up singing three different anthems . . . Britain's "God Save the Queen," Japan's "Kimigayo," and Malaysia's "Negara Ku" . . . before coming up with their own.)

Meanwhile an initially far richer Malaysia favored ethnic Malays first, not talent.

Many well-educated Chinese and Indians left.

And when the country's leader was warned, in the mid-1970s . . .

That some of the most educated and entrepreneurial were leaving . . .

He responded . . .

"This is not a 'brains drain.' It is a 'trouble drain' . . . It drains trouble out of Malaysia."[2]

Today Singaporeans have a standard of living comparable to that of the United States . . .

Almost three times higher than Malaysia.

MALAYSIA THREW SINGAPORE OUT IN 1965 . . . TODAY . . .

	Wealth per person	International reserves (per person)
MALAYSIA	51 in world	$2,131
SINGAPORE	16 in world	$21,405

(Malaysia's current leader is trying to reverse these trends with massive investments in telecommunications, airports, highways, and cyber cities . . . But he is running out of time, capital, and political leeway.)

The small can do very well given open markets, smart people, and relative peace.

SMALL AND RICH . . . LARGE AND POOR . . .[3]

SMALL, NATURAL-RESOURCE-POOR	REAL WEALTH GENERATED PER PERSON ($2002)	LARGE, RESOURCE-RICH	REAL WEALTH GENERATED PER PERSON ($2002)
Taiwan	13,000	Nigeria	250
Israel	19,000	India	500
Andorra	16,600	Angola	600
Liechtenstein	25,000	Congo	700
Hong Kong	25,500	China	950
Monaco	27,000	Indonesia	1,100
Singapore	27,000	Kazakhstan	1,900
Bahamas	30,000	Colombia	2,300
Belgium	31,000	Venezuela	2,300
Netherlands	31,000	Mexico	3,700
Iceland	31,500	South Africa	4,000
Denmark	40,000	Brazil	4,600
Switzerland	47,000	Saudi Arabia	7,600
Luxembourg	59,000		

Of course there are some successful large and resource-rich countries . . . Like the United States and Canada . . . But these were not poor countries during the first decades of the twentieth century . . . When the knowledge economy began emerging.

IN OTHER WORDS, THEY USED THEIR RESOURCES AND DEVELOPED WHEN THE RULES WERE DIFFERENT.

(Oh, and by the way, before anyone gets too complacent . . . According to the World Bank, the United States is not the world's richest country on a per-person basis; it ranked tenth in 1998. Canada ranked twenty-sixth. Countries like Switzerland, Singapore, and Finland have done a better job at integrating different cultures and languages within the same border, educating high school students, and distributing wealth. It is a few entrepreneurs, many of them foreign-born, who keep the United States on top.)

HOW CAN COUNTRIES
RICH IN NATURAL
RESOURCES
GET SO POOR
DURING THIS CENTURY?

Many scholars have focused on **corruption and mismanagement** . . .

But this is only part of the story.

Natural resources provide a temptation . . .

To **exploit** them first . . .

Accumulate capital . . .

And then *educate* people.

But this logic is nonfunctional in a global economy where . . .

We produce more . . . for less.

The global market is flooded with grain, milk, steel, plastic, oil . . .

From 1845 to 1998, the basic price of commodities fluctuated significantly . . .

But the trend has been unrelenting.

The average commodity is worth

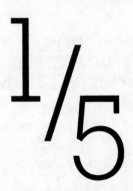

1/5

of what it was a century and a half ago.

This is an old story. In 1600 one of the most valuable pieces of real estate in the world was Run . . . Which produced nutmeg. The three great naval powers, the English, the Dutch, and the Portuguese, fought bitterly to control this tiny island. In 1667 the Dutch won and the English, to save face, asked the Dutch for a useless property . . . New Amsterdam . . .

Today Manhattan.

(Now Run does not even appear in most atlases.)[4]

COMMODITY PRICE INDEX 1845–1999[5]

(1845–50 = 100)

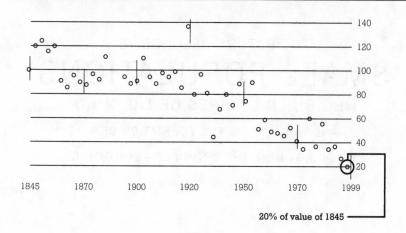

20% of value of 1845

You can see the effects of cheap commodities in Aberdeen, Scotland . . .
(Still nicknamed Granite City because of its great quarries).
When the city council recently decided to restore various buildings . . .
It chose not to use local stone.
Shipping stone from China . . .
A mere 10,000 kilometers away . . .
Was cheaper.

Shipping coals to Newcastle used to be a joke.
Today it is deadly serious.
Many countries fail to realize the implications of the
"Singapore Law":

THE FUTURE BELONGS TO
SMALL POPULATIONS
WHO BUILD EMPIRES OF THE MIND,
AND WHO IGNORE THE TEMPTATION OF—
OR DO NOT HAVE THE OPTION OF—EXPLOITING
NATURAL RESOURCES.

Tiny Brunei . . .

On the island of Borneo . . .[6]

Very different from Singapore.

The current sultan of Brunei was once thought the world's richest man . . .
(Until his brother allegedly absconded with or misspent billions.)

Able to spend two billion on the world's largest inhabited palace . . .
(A space larger than the Vatican . . . run by Hyatt.)

But despite a wildly generous "Shellfare" state . . .
(Financed by Shell and other oil companies.)

The country is unlikely to succeed because over 75 percent of its workers are government bureaucrats or live off government contracts.

And its few successful "entrepreneurs" do not make things or think things . . .

They import BMWs and other toys.

As long as formal *Titles* matter more than brains . . .

Those who dominate are those who choose the right parents . . . and grandparents . . .

Not those who work or study harder.

There is little hope of lasting wealth . . .

When oil runs out . . .

There will be some good ruins.

(A century from now, tourists may scramble over the ruins of Brunei's billion-dollar Empire Hotel and Country Club, which hosted the 2000 Asia-Pacific Economic Cooperation forum. Come to think of it, they can do so today. Large parts of this gold-plated hotel, designed to impress the dozens of world leaders who attended, are unlikely to ever get finished, if anyone has any economic sense.)

AND A LOT OF PEOPLE
WHO WERE WEALTHY ONCE UPON
A TIME . . . WILL ONCE AGAIN BECOME VERY POOR . . .

(just following the example of other oil powers . . . say, the Iranians, Iraqis, Saudis, Venezuelans . . .)

Countries whose economies remain natural-resource-based . . .

Have to produce more and more . . .

To earn the same.

As their populations grow . . .

Most of these countries get poorer and poorer.

A farmer has to produce three times as much grain . . .

To make the same income . . .

He had fifty years ago . . .

(Without taking inflation into account.)

So farms get bigger . . .

And employ fewer people.

Ideas are different . . .

The more you spread them . . .

The more valuable they become.

So now those who produce knowledge . . .

Get consistently richer.

And that is why the great empires of the twenty-first century are . . .
And will remain . . .

[EMPIRES OF THE MIND.]

Churchill was right . . .
, (No wonder he so often looked smug.)

When societies turn their back on technology and stop building empires of the mind . . . They decline rapidly. Look at the great Muslim states. The reason we know so much about Greek culture today is because Muslim scholars translated and preserved key manuscripts, maps, plays, science experiments. In A.D. 900, the world's two leading cities were Baghdad and Cordoba (in Spain) . . . There were no public libraries throughout Christian Europe . . . Moorish Spain had seventeen . . . One of these had more than 400,000 books. *Algebra* and *chemistry* are Arabic words . . . We still use Arabic numbers today. Scholars focused on animal husbandry, botany, zoology, and psychology. Meanwhile, Rome's great Colosseum was an apartment house invaded by squatters.[7] Alas, neither Rome, nor Andalusia, nor Baghdad remain empires . . . Great scholars now choose to build empires of the mind elsewhere.

V

DATA DRIVES EMPIRES

Perhaps there is no better example . . .
Of the power knowledge has . . .
To rapidly transform an economy . . .
Than the digital revolution.

The digital revolution transformed not just computer companies . . .
But also . . .
Television . . .
Cable . . .
Pagers . . .
Radio . . .
Newspapers . . .
Magazines . . .
Telephones . . .
Photography . . . (U.S. consumers buy far more digital cameras than film cameras.)[1]

It did so **by creating and spreading a new language . . .**

And language is a powerful medium.

WHAT DISTINGUISHES PEOPLE FROM ANIMALS IS THE ABILITY TO UNDERSTAND ABSTRACT CONCEPTS AND TO COMMUNICATE THESE CONCEPTS.

The history of civilization can be summarized as a series of efforts to transmit and use increasing amounts of information.

There are many ways to do this.

Stories from grandparents . . .
Songs and anthems . . .
Photographs . . .
Newspapers . . .
Television . . .
Paintings . . .
Books . . .

As we get more information . . .

We get better at processing and retransmitting information.

We can see how this process occurs through the evolution of language.

When humans lived in caves . . .

If someone wanted to tell the group an elk had gone by, and it was time to hunt . . .

They would point to part of a drawing on the cave wall.

Each thought equaled a drawing.

But **it was very hard to transmit abstract concepts . . .**

Transmit a lot of information quickly . . .

Or standardize communication and language.

(As late as 1788, Australia's aboriginal population had around 270 separate and distinct languages and more than 600 dialects, some highly sophisticated.)

Basic written alphabets (created with cuneiform letters and hieroglyphs) supported the first great civilizations in Mesopotamia (Iraq) and Egypt.[2]

The Chinese stylized primitive drawings and languages into simplified characters . . .

Began to standardize sounds and words . . .

Combined them to express abstract thoughts . . .

And launched one of the world's great civilizations.

(By 930, they were printing thousands of books using woodblocks.)

The Japanese call these letters Kanji . . .

And can communicate any abstract or technological concept through strings of characters.

"Tiger" in Chinese . . . from cave to present . . .

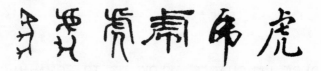

Although a little simpler . . .

This was not a purely abstract alphabet.

So to read a basic newspaper, or to type a letter, students have to memorize close to 10,000 basic characters . . .

And keep learning, and memorizing, throughout their lives.

The Greeks simplified all concepts into a few letters . . .

Which, combined in different ways, could express almost any concept . . .

And set the basis for the twenty-six letters we use today.

A, B, C, . . .

We no longer require drawings to communicate the most abstract of concepts.

And, ironically, fewer symbols led to more variety and complexity . . .

Japanese has around 120 possible syllables . . .

English has more than 1,000.

MODERN LANGUAGES ALLOW US TO COMMUNICATE A LOT OF DATA REASONABLY QUICKLY, BUT THIS IS NOT WHERE THE STORY ENDS.

Over the last few decades . . .

Much of the world . . .

Is communicating through a far . . .

Simpler alphabet.

In the mid-twentieth century, engineers invented a device to transmit, or not transmit, electrical impulses.

They called it a transistor . . .

And it led to the first computers.

Computers were stupid . . .

They could not comprehend or read letters, images, or music.

The only thing they could understand was a binary language . . .

On/off . . . Electric pulse/no pulse . . . Light/no light.

This required creating a two-letter alphabet . . .

To catalog, transmit, and process all human knowledge.

If an electrical impulse goes through one of the millions of transistors embedded in a computer's microchips . . . the computer reads "1" . . .

If there is no impulse, it reads "0."

TODAY'S DOMINANT ALPHABET NO LONGER CODES
IN TWENTY-SIX LETTERS BUT IN TWO . . .[3]

1s AND 0s

THE DIGITAL ALPHABET
ENCODES AND TRANSMITS INFORMATION . . .
WITH EXTRAORDINARY SPEED AND ACCURACY . . .
AND IT HAS BECOME THE WORLD'S MAIN LANGUAGE.

In 1997, telephone wires, for the first time, carried more digital
data than voice conversations.

By 2003, less than 3 percent of the data transmitted across
telephone networks is expected to be a conversation between one
person and another.[4]

> In a digital world . . .
>
> When you pick up the phone and dial . . .
>
> The phone becomes a part of a digital network . . .
>
> Which means your voice is no longer transmitted as
> sound waves . . .
>
> But instead as a long string of 1s and 0s.

For example, your phone may transmit:

01001000 01101001 00100000 01101101 01101111 01101101

And the person who receives these digital strings hears . . .

"Hi, Mom."

(It is actually far more complex than simply writing out these ASCII characters, because the information also has to transmit tone, volume, pauses . . .
But you get the general idea.)

Change a few ones and zeroes:

01000010 01111001 01100101 00100000 01101101 01101111
01101101

And you get, "Bye, Mom."

As we got better at coding.

The digital alphabet began to transmit . . .

Music . . .

Photos . . .

Movies . . .

Virtual realities.

(Music is no longer coded onto vinyl records with varying ridges read by a needle; instead we use mirror-surfaced CDs that contain only 1s and 0s.)

The speed of data transmission is getting faster and faster.

We now use fiber optics to transmit information . . .

If a light pulse goes through a computer it reads 1 . . .

If not it reads 0.

This allows information to be coded, transmitted, and decoded very quickly.

In 1999, Bell Labs scientists were able to transmit

1,600,000,000,000

bits of information (1s and 0s) . . . in one second . . . across a single fiber-optic line . . . no wider than one of your hairs.

In other words . . .

You could transmit all U.S. telephone conversations . . .

During the peak hour . . .

Of the peak day . . .

Across a single fiber-optic cable . . .[5]

(Or the entire contents of the Library of Congress in about six seconds.)

When people share a common language . . .
Regions often come together . . .
So too do companies.

*Two decades ago, the languages spoken at a
newspaper, television studio, paging service, telephone
company, recording studio, film studio . . . were
each quite different.*

Today these companies all use . . .
And exchange . . .
Digital information.

So it should shock no one . . .
To see that a common language . . .
Led to a series of mergers . . .

And
created
the
largest
industrial
conglomerates . . .

The world has ever seen.

JANUARY 2000
BIRTHED AN UNPRECEDENTED BEHEMOTH . . .

What started close to eight decades ago as a
newsmagazine, *Time* . . .
Had already gone through a series of transformations:
Merging with a large film company (Warner) in 1990 . . .
And with the largest cable-news operation (CNN) in 1995 . . .
To create the world's most powerful media company with . . .
Thirty magazines and close to 120 million readers . . .
More than 3,000 movies . . .
And a cable-news operation that could reach up to
one billion people.

But the largest news-entertainment gorilla . . .
Was eaten up by a company that was founded in 1985 . . .
America Online.

(At the end of 1997, Time Warner's assets were about forty times those of
AOL . . . But the Internet boom changed things . . . So did AOL's 27 million
subscribers.)

The speed and scope of technological change is such that . . .
AOL became twice as valuable as Time Warner . . .
And became the lead company . . .
In the **world's largest merger.**

(A new company that expects to be Microsoft's main competitor . . . By
January 2001 AOL Time Warner controlled 33 percent of Americans' time on
the Net . . . Yahoo! had 7 percent . . . Microsoft 6 percent.)

Things change very quickly in a digital world.

Even companies . . .

That have more cash . . .

That have more information . . .

And that have become larger . . .

Than most national economies . . .

Are by no means guaranteed survival . . .

Either they get smarter . . .

And grow . . .

Or someone will eat them up.[6]

THE POWER OF TECHNOLOGY . . .

TO BUILD . . .

AND DESTROY . . .

IS SUCH . . .

THAT IT IS LIKELY . . .

FEW OF US HAVE EVER HEARD . . .

THE NAME OF WHAT WILL BE THE

WORLD'S LARGEST COMPANY . . .

IN 2020.

The digital revolution..................................is just the beginning.

Because a new, far more powerful revolution is brewing . . .

One that will change most life forms on the planet . . .

Literally.

VI

GENETICS . . .
THE NEXT
DOMINANT
LANGUAGE

*Perhaps the most important discovery of the
twentieth century was to learn to identify and read
the code of life.*

*And perhaps the most important challenge we
will face in the twenty-first century . . . is
how . . . and when . . . to apply this knowledge.*

(Harvard, with its classic modesty, has located its biochemistry, genomics, and molecular-biology labs just off
Divinity Avenue.)[1]

It took nearly a century and a half to start to read the language that determines all life processes.

In the 1850s, an Austrian monk,

Gregor Mendel,

began experimenting in the garden of his monastery.[2]

He used the pollen of some plants to carefully fertilize other plants . . . Mostly peas.

By carrying out these experiments deliberately and carefully recording the results, Mendel was able to observe that various traits present in grandparents, mothers, and fathers could be passed on to offspring . . .

And he could catalog which traits tended to dominate . . .

Thus giving rise to a new discipline . . .

Which we now call genetics.

But no one paid much attention to Mendel's discoveries until almost a half century after he began experimenting (sixteen years after his death).

Over the course of the twentieth century . . .

Scientists gradually realized the importance of Mendel's work . . .

And began conducting their own experiments . . .

And drawing their own conclusions . . .

For *better* . . . or worse.

Doctors noted that some diseases attacked some grandparents, children, and grandchildren.
Disease could be inherited . . .
And could sometimes be statistically predicted . . .
So today, many genetic diseases can be prevented, or treated.

But there were also those who used genetics to create a pseudoscience . . .

Eugenics.

They drew the wrong conclusions . . . and used them to justify racist laws.

They believed there were superior . . .
And inferior . . .
Races.

(A wonderful book titled *The Mismeasure of Man* shows how even "objective scientists" could misreport measures of cranial size if they expected to find racial differences.)[3]

Hitler loved eugenics . . .
He used the concept to justify . . .
The murder of millions . . .
Because they were not part of a "chosen race."

(Then again . . . from 1901 through 2000, the constitution of Alabama read, "The Legislature shall never pass any law to authorize or legalize any marriage between any white person and a Negro, or descendant of a Negro . . ." This law was finally abolished through a referendum in November 2000. Still, 40 percent of voters . . . over 544,000 people . . . voted against this change.)

(P.S. Alabama is not alone . . . In 2000, South Carolina's government was racked by a vigorous debate because the Confederate flag was still flying over the State House, and many African Americans considered this an insult . . . State senator Arthur Ravenel responded by referring to the NAACP as the "National Association of Retarded People" . . .)

Genetics is a very powerful instrument . . .

So we all have a stake . . .

And a reason . . .

To understand . . .

And participate . . .

In debates . . .

As to what we should . . .

And should not do . . .

As we study and alter . . .

The code that governs . . .

All life on the planet.

(Not that this is without controversy . . . There is a growing movement, particularly in Europe, against genetic engineering . . . Various organizations and individuals have taken a strong stance.)[4]

Genetic knowledge helps keep most of us alive.

In 1804 there were one billion people on the planet . . . 1927 two billion . . . 1999 six billion . . .

We would have **starved** long ago . . .

If agricultural productivity had not increased much faster than population.

The grains and animals we eat today . . . are "unnatural."

We have been modifying the genetics of foods for a little while.

(Although we have been around for close to 1.8 million years . . . we started systematic cultivation only in the last 11,000 years.)

Most foods we eat today are the result of careful breeding and cultivation . . .

Particularly of a very few species.

Of over 200,000 plants, only a dozen account for 80 percent of all crops.[5]

Many of today's staples were once inedible or hard to eat . . .

"Natural" tomatoes are small green berries . . .[6]

WILD PERUVIAN TOMATOES . . .

(L. hirsutum)

Are small . . . Green . . . Slightly poisonous.

Kidney beans can be poisonous if undercooked . . .[7]

"Natural" corncobs are . . . the size of your fingernail . . . Covered with irregular . . . multicolored kernels . . . that drop off easily.

Most of the flowers and fruits in our houses are artificial hybrids . . .

And "natural" dogs . . . are wolves.

Most genetic "engineering," so far, has been haphazard . . .
Through 1953, we had little idea how heredity was coded . . .
Much less how to read it in detail.

Then a young scientist, James Watson, and an orthodox British professor,
Francis Crick, discovered that the traits we inherit from our ancestors
depend on a complex molecule with an unpronounceable name . . .

DEOXYRIBONUCLEIC ACID.

(We know this compound by its shorthand . . . **DNA**.)[8]

 DNA contains the code of all life processes . . .[9]
Bacteria, worms, fish, birds . . . Humans . . .

Each plant, animal, or bacterium carries its entire genetic code . . .

Inside almost every one of its cells.

The structure of each molecule of DNA is like a ladder . . .

The sides of the ladder are built with sugar and phosphate.

Four substances—adenine, thymine, cytosine, and guanine—
form the rungs of the ladder.

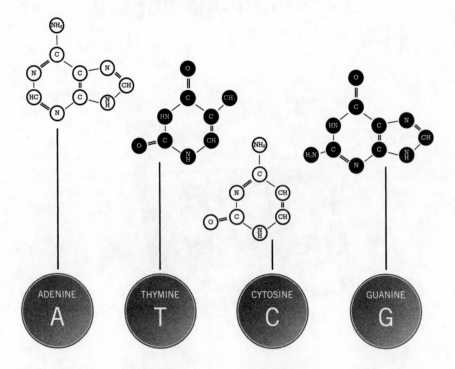

Because DNA is a very large molecule . . .

And it would be endless to write out time and again
the full name . . .

Of each base pair that forms a rung on the ladder . . .

Scientists simplify their notations into A, T, C, and G.

In complex animals . . .

Like humans . . .

The amount of information carried and processed . . .

To preserve life functions is staggering.

If we stretched out the DNA contained in each of our cells . . .
It would be about six feet long (. . . but it folds up into
trillionths of an inch).

OUR GENETIC CODE CONSISTS OF

THREE BILLION LETTERS

(A, T, C, and G) . . .

AND THIS CODE IS REPLICATED TWICE WITHIN . . .
EACH ONE . . . OF OUR

HUNDRED TRILLION CELLS.

(But before you get too excited by "this wondrous machine called
man" . . . Remember that an ear of corn has 2.6 billion base pairs, and
wheat more than 16 billion.)

By 2000 we were able . . .

To read the entire genetic code of a human being . . .

Otherwise known as the human genome.

Writing down on paper a copy of the genome we carry within
each of our cells . . .

Would require the equivalent of 248 Manhattan phone
directories . . .

A typical paragraph might read:

```
AAATTCCTTTAGGGATTTAGGCCCCTGAGAAAATCCGGCCC-
CGCCTCCTAGAGATCTCGATATTTAGGGGGATTTGGGGCC-
CATTTTAGGGGGATTATTATATAAATACCCCTATATATAAAAAGAG
GATTTTCTTCCATACTTTTCCTCCAAAATTTTGGGGGCCCAT-
TATATTTTAAATACTTCCCTGTAATGTTGGAGGAAATATTTG-
TAAATATAATATAAAGAGATTTATATATTAGAGAGAGGCCAG . . .
```

IF SOMEONE SPENT HER ENTIRE LIFE READING A COPY OF ONE PERSON'S GENOME . . . SHE WOULD BARELY FINISH . . . MUCH LESS UNDERSTAND . . . OR REMEMBER . . . WHAT SHE READ.

But it is important to know what one is reading.

A word can mean very different things depending on . . .

The **letters used** . . .

The **order of the letters** . . .

And **where these letters are placed** within a sentence.

Similar letters can have very different meanings . . .

Polish the Polish furniture.

Even the same letters can have different meanings . . .

"letsgotogether" can mean "let's go together" or "let's go to get her."[10]

The same is true of a computer program.

If one writes 00101111, the computer divides two numbers, whereas if one writes 00101010, it multiplies.

And the same is true of DNA.

If science or nature substitutes one string of ATCGs for another (particularly within a section of DNA that codes actively),

The organism may be drastically altered.[11]

Small variations within a genetic code can have
enormous impact.

We have only one-third more genes than a worm . . .

Around 99 percent of your genes have partial counterparts
in mice . . .

(No wonder some people act like rats.)

So scientists test various medicines and procedures on
mice and other animals . . .

Before attempting to treat humans.

*By the time we get to bonobos (pygmy chimps), the
gene overlap with humans is about 98.5%.*[12]

(Something that becomes obvious when one deals with certain politicians . . . Not surprising, given that we began speciating from the great apes
about seven million years ago . . . and began giving up a nomadic existence
only in the past 13,000 years.)

*The variation between your genome and that of your brother, neighbor,
worst enemy, or that of any other person on the planet is minuscule.*

**Of every thousand letters of genetic code, we differ from our neigh-
bor by less than one letter . . . That is, by less than one A, T, C, or G
base pair.**

And because only a few regions of our total DNA actively code
life functions . . .

*The real differences between one person and another are less
than 0.0003 percent.*

(Chimps have a more varied genome, because they have been around a lot
longer than humans. Two animals taken at random from Africa have at least
four times more mutations within their genomes than humans.)[13]

But minute changes in a genome are enough to make some
tall, others short, some fat, some slim, some blond, others
redheads.

They are also **enough to make some healthy and others very
sick.**

A few letters' difference can make a difference in who lives
a long time . . .

And who does not.

Some diseases attack everyone who lives in the same
environment . . .

Others do not.

Sometimes only one child, or a few grandchildren, or one in five
neighbors, comes down with a specific disease.

Often the difference in whether one gets cancer or not . . .

How violent the cancer is . . .

Depends on minute variations of the letters within our genetic code.[14]

For example, women who carry a specific genetic variation known as BRCA-1 are seven times more likely to get breast cancer. And this cancer is likely to occur earlier in life and be more malignant.[15]

Even three letters out of three billion can make a huge difference.

If you are missing CTT at a specific spot in your genome . . .

You carry cystic fibrosis.[16]

(But beware . . . Genes are not necessarily destiny. Eric Lander, head of MIT's Whitehead Institute, likes to point out that there has been enormous variation in how humans govern themselves, make money, think about ethics or war . . . even though there has been very little variation in the human genome. Your environment and the choices you make are at least as important as your genes.)

As we get smarter about why one person gets sick and another does not . . .

And how to prevent and cure certain diseases.

The consequences are significant.

A child born in 1900 in the United States had a life expectancy of 46 years . . .

By the end of the century it was 78 . . .

Close to a 70 percent increase . . .

And **genetics is likely to increase this life span significantly.**

The types of diseases people die from today are quite different from those of a century ago.

(In 1900, the three leading causes of death in the United States were pneumonia, tuberculosis, and diarrhea. Today, they are heart disease, cancer, and stroke.)

One rarely sees mass epidemics (besides AIDS and maybe avian flu) . . .

Instead, more and more people fall victim to gradual degenerative diseases . . .

Like arteriosclerosis, Alzheimer's, and cancer.

If, how, and when one gets these diseases depends in part on genes . . .

And in part on one's lifestyle.

Smoking, for instance, can trigger gene mutations and lead to cancer.[17]

How long you . . .

Your children . . .

And your grandchildren . . .

Live . . .

Depends increasingly . . .

On understanding the human genome . . .

And on understanding . . .

The genetic code of the organisms that attack our bodies.

(Substituting a single G for an A inside the three-million-letter malaria code could kill you . . . Substituting a single amino acid inside one of your proteins can lead to sleep deprivation.)[18]

Bill Haseltine, former CEO of Human Genome Sciences . . .

Is a smart, tough, controversial scientist . . .

Who is absolutely determined to prove . . .

That our body contains a medicine chest . . .

Full of substances that heal wounds, fight infections, mend broken tissue.

Haseltine believes that understanding the genes that allow us to heal ourselves . . .

Will provide the most powerful medicines in the future . . .

And will allow us to live 120 years . . .

Before we start regenerating our own tissues . . .

After that, we may be "young and healthy forever."[19]

(Haseltine has never been accused of shyness . . . nor have many of the scientists leading the gene revolution . . . Perhaps this is why *The Scientist* wrote that the genome is an anagram for "ego men.")

But before you dismiss Haseltine as another optimistic quack . . .

Or assume this is just science fiction . . .

You might consider that . . .

Haseltine was one of the United States' leading AIDS researchers . . .

CEO of a major pharmaceutical company . . .

(And you might also consider that Haseltine's previous company applied for 9,200 patents.)

Remember that in developed countries today . . .

Communicable diseases kill only 15 percent of the population.[20]

Many women are expected to live ninety years . . .

Because they have fewer accidents and are less prone to fight than men.

And that many of us can hope to live to be over 100 . . .

Something that would have been almost unthinkable at the beginning of the twentieth century.

Sarah Knauss used to have an unusual family photograph on her dresser . . .
She was surrounded by her family . . .
A ninety-six-year-old daughter . . .
Her grandson (seventy-five) . . .
Great-grandson (forty-six) . . .
Great-great-grandson (twenty-four) . . .
And a three-year-old great-great-great-grandson . . .
Sarah died when she was 119.

These photos may become more common . . .
In September 2000, *Science* reported that two drugs, antioxidants . . .
Could increase the life span of nematode worms by 50 percent . . .
The first example of a drug used to slow aging.[21]

Our children will likely . . .
Be running on the beach well into their nineties . . .
Powered by replacement body parts . . .
That come from a variety of people and are grown in various countries.

The ability to understand and cope with basic public-health issues . . .
Makes a critical difference in a country's future.

HOW THE WORLD LIVES AND DIES[22]
(disability-adjusted life years)

1900	1990	2020
Communicable	Flu	Heart disease
diseases	Diarrhea	Depression
were the	Pregnancy problems	Road accidents
leading cause	Depression	Strokes
of death	Heart disease	Chronic pulmonary disease
		Chronic non-communicable diseases will cause 73% of deaths

Some countries have done a good job improving basic health care.

Between 1982 and 2000, China's population increased 19 percent . . .

But the number of people over sixty increased 72 percent.

MEANWHILE, IN ZIMBABWE . . .
THEY TRAIN THREE EXECUTIVES FOR EACH JOB . . .
BECAUSE TWO MAY DIE . . .
OF AIDS.

SOCIETIES AND PEOPLES WHO UNDERSTAND THE GENETIC ALPHABET . . .

ARE LIKELY TO LIVE LONGER . . .

AND GET RICHER.

(Sounds simple . . .But remember what Monty Python sort of said . . . A dictionary has all the words contained in every work of literature . . . Therefore all Shakespeare had to do was to get the right number of words . . . in the right order.)

But most societies do not understand genetic discovery . . .

Or the challenges that arise from these discoveries . . .

And that makes them, for all practical purposes . . .

Functionally illiterate . . .

In the language that codes all life on this planet.

VII

GENETICS IS . . . A HOCKEY STICK

["WE ARE LEARNING THE
LANGUAGE IN WHICH GOD
CREATED LIFE . . . WITHOUT
A DOUBT, THIS IS THE MOST
IMPORTANT, MOST WONDROUS
MAP EVER PRODUCED
BY HUMANKIND."]

PRESIDENT CLINTON
announcing the assembly
of the human genome sequence
JUNE 26, 2000

An extraordinary thing happened in 1995 . . .

Robert Fleishmann and Craig Venter published the first full genome of a living organism . . .[1]

An obscure bacterium called *Haemophilus influenzae* . . . (which can give you meningitis).[2]

For the first time, scientists could see a complete map of how a life was programmed.[3]

THE FIRST GENETIC MAP OF A LIVING ORGANISM . . .
1,743 GENES . . .
(40% UNKNOWN)
1,830,137 BASE PAIRS

Short term . . .
This map allowed doctors to treat meningitis and ear infections more effectively . . .
And unleashed new ways to map and combat diseases.

(By year's end [2001] we should be able to study the entire operating code of over one hundred diseases including cholera, malaria, anthrax, pneumonia, Legionnaires', plague, gonorrhea, tuberculosis . . .)[4]

Long term . . .

It marked the beginning of a new era in mankind's history.

Over the past five decades, we have started to accumulate the instruments and knowledge required . . .

To control . . .

Directly and deliberately . . .

The evolution of our species . . .

And that of every other species on the planet.

What Venter achieved culminated centuries of efforts to understand heredity and the transmission of biological traits . . .

It changed the way in which biology is defined and studied.

Centuries ago, biologists began systematically observing and experimenting on live organisms.

In Latin . . . **biology was focused on the** in vivo **study of organisms . . .**

(Or almost in vivo—Leonardo da Vinci used to dissect stolen cadavers.)

After the initial great discoveries . . .

Most biology moved indoors . . .

Into small beakers, test tubes, petri dishes, and other glass containers . . .

Life was studied mostly in vitro.

Venter began combining the power of a lab with that of a computer.

As silicon chips and automated machines combined with breakthrough biology . . .

We saw the **emergence of** in silico biology.

This transition from in vivo to in vitro to in silico is in some ways similar to the road taken by physics.

Newton's falling apples gave way to . . .

Experiments on physical objects, which gradually gave way to complex labs and accelerators.

Meanwhile, a parallel, equally powerful discipline, theoretical physics, also emerged . . .

And began predicting what labs would find . . . in years or decades.

(Astronomy has also become a discipline in which theory usually precedes visual or experimental confirmation . . . This started after an English mathematician observed Uranus' orbital behavior and predicted the discovery of a new planet, Neptune, in 1841.)[5]

Even extraordinary scientists like James Watson (co-discoverer of DNA) . . .

Did not understand . . .

And in fact disdained . . .

What Venter was doing . . .

Because they failed to understand . . .

The extraordinary power of the digital revolution.

By bringing together standardization, mechanization, and supercomputers, Venter was able to unleash an avalanche of knowledge.[6]

(For those skeptics out there who doubt just how far computers, data processing, and visual imaging have come, consider the following . . . Super Bowl XXXV (2001) was covered with three-dimensional, 180-degree replay cameras . . . But you cannot watch the 1967 Super Bowl . . . There is no complete film in existence.)

In 1991, all of the world's labs had identified only 2,000 genes . . .

(That is, the discrete instructions that exist within our genome.)

By 1995, Venter and his team were able to publish a partial map of 35,000 genes.[7]

(Some of which ended up being duplicates, but still quite a coup.)

By 2000, Venter's company, Celera, had accumulated gene-data equivalent of six Libraries of Congress in its basement computer.

There is little evidence that the overwhelming volume of new knowledge will slow . . .

Because research budgets are increasing massively . . .

And the cost of sequencing is dropping.

In 1974, Monsanto estimated that sequencing a single gene would cost the equivalent of . . .

$150,000,000 . . .

By 1998, the cost of sequencing a gene was . . . $150.

By 2001 it could be less than . . . $50.

(The "average gene" is about 27,894 base pairs . . . A,T,C,Gs . . . Say, eight pages of tightly packed footnotes.)

As costs collapsed, knowledge, and patents, exploded.

During the 1952 debate on patent reform, Congressman P. J. Federico wrote:

Patentable subject matter "includes anything under the sun made by man."

In 1971, a General Electric scientist attempted to patent a bacterium . . .

This led to a series of complex court disputes . . .

Which culminated in a divided (5–4) Supreme Court decision . . .

That allowed patents on living organisms and genes.[8]

BY 1988, SCIENTISTS WERE PATENTING NOT JUST BACTERIA AND PLANTS . . . BUT ENTIRE ANIMALS.

An extraordinary doctor, Phil Leder, created the "Harvard mouse" . . . An animal that was highly susceptible to cancer . . . And that could be used to test different treatments . . . in various labs. This mouse has saved . . . or prolonged . . . millions of lives.

HARVARD
MOUSE

PHIL LEDER

Soon the volume of data and new gene sequences became overwhelming.

As station wagons and trucks disgorged boxes of data and requests from various and sundry labs . . .

The poor folks at the U.S. Patent and Trademark Office (USPTO) basically gave up on seeing the light of day or taking vacations . . .

In **1991,** the USPTO received an **overwhelming 4,000 requests for patents** on various expressed sequence tags (ESTs).

GENETIC PATENT REQUESTS
(ESTs 1991)

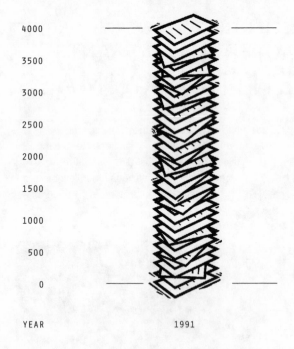

4000	
3500	
3000	
2500	
2000	
1500	
1000	
500	
0	

YEAR   1991

Some at the Patent Office thought this was simply the first wave of discovery and that soon things would calm down . . .
They were wrong.

By 1995, EST patent requests increased to 22,000 per year.

GENETIC PATENT REQUESTS
(ESTs 1991, 1995)

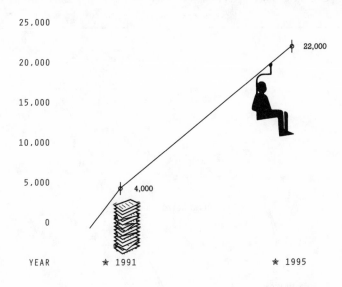

And in the bowels of the Patent Office, you could hear *@&%#!!!! . . .

BUT AS BATMAN LIKED TO SAY . . .

THE WORST IS YET
TO COME . . .

One year later, in 1995 . . .

EST patent requests increased to an unmanageable
500,000 per year (bundled into single patent requests, thousands of pages long).

GENETIC PATENT REQUESTS
(ESTs 1991–96)

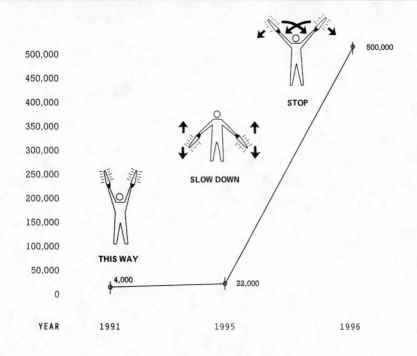

And the Patent Office
was forced
to change
the rules
as to how one could apply
for basic genetic patents.[9]

More than 100 whole animals have been patented.

This is just the beginning . . .

Scientists are discovering, creating, modifying tens of thousands of life forms.

(Someday, we may also revive extinct species. Australian scientists hope to use the DNA of a Tasmanian tiger to make the animal's disappearance from the planet "a 70-year hiccup." Apparently some, besides Steven Spielberg, already dream of woolly mammoths . . . This is a good thing, given that 11 percent of all birds, 25 percent of mammals, and 34 percent of fish are threatened with extinction.)[10]

Patent statistics reflect just part of the change.

Scientists are overwhelmed by new knowledge . . .

Every week . . .

Every day . . .

If you graph this knowledge . . .

It is not a straight upward curve . . .

Or a gradual curve . . .

GENETIC KNOWLEDGE IS A HOCKEY STICK . . .
IT BOUNCES ALONG THE BOTTOM OF THE GRAPH . . .
FOR DECADES . . .
AND SUDDENLY EXPLODES
UPWARD.

(This is not unique. There are many databases that grow on an exponential instead of a linear basis, including computation, communication, brain scans, miniaturization . . . Each growing database accelerates change in the others.)[11]

You see this trend in almost any genetic data graph . . .

Whether it is nucleotide sequences . . .

Proteins . . .

Genome maps . . .

Growth has exploded.

EXPLOSIVE GROWTH IN BIO DATA[12]

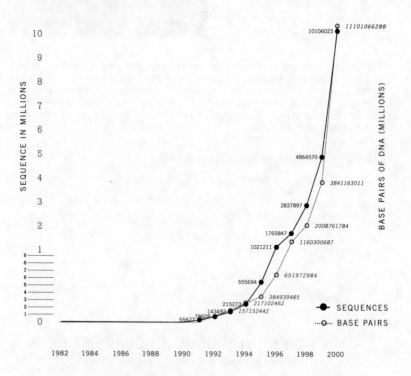

In 1980, it took a year to assemble 12,000 DNA base pairs;

In 1996, it took twenty minutes . . .

At the end of 1999, it took less than a minute . . .

Toward the end of 2004 . . .

The Venter lab could sequence one hundred million
letters of new gene code per day.

Researchers have accumulated enough sequencers and computers . . .
To be able to sequence every major disease . . .
In a single day . . .
And someday, given enough scientists and resources . . .
They might just do so.

But microbial genomes are trivial compared with the
complexity of . . .

Flies . . .

Mustard weed . . .

Humans . . .

Three genomes were completed by 2001.

LANDMARKS IN FULLY SEQUENCED ORGANISMS . . .

ORGANISM	KNOWN AS	DNA BASE PAIRS (MILLIONS)	COMPLETED
Haemophilus influenzae	Meningitis	2	1995
Caenorhabditis elegans	Worm	100	1998
Arabidopsis thaliana	Mustard weed	119	2000
Drosophila melanogaster	Fruit fly	180	2000
Homo sapiens	Human	3,100	2001

Just as having a geographic map of a coast, river, or range allows us to explore, navigate, and colonize . . .

Having the gene maps of basic life forms will allow us to begin to understand them . . .

A LOT FASTER.

PHARMACEUTICALS, ENERGY, COSMETICS, CHEMICALS, AGRIBUSINESS . . .

BUSINESSES THAT USED TO BE PATENT-DRIVEN . . .

ARE NOW OVERWHELMED BY RAPID . . .

DATA CYCLES.

(AND THOSE WHO MISS A CYCLE HAD BETTER BE WILLING TO RUN FAST.)

Political parentheses . . .
It's not just CEOs who are discombobulated by change .
When Bill and Hillary went to Italy in the late 1980s . . .
They met some Italian Communists . . .
Who were anti-Soviet, pro-NATO, and pro-free enterprise . . .
Clinton thought: "I've got to be very careful about what words
mean anymore."

    CEOs who think that they are doing well because . . .

    They launch one or two major drugs per year . . .

    And can rest on their laurels until the medicine goes off patent in
    a couple of decades . . .

    Are in for a nasty surprise.

Pharmaceuticals are likely to become niche products . . .

Targeted to individual genotypes . . .

And new products will likely appear weekly.

(This is going to create more pressure on M.D.s and pharmacists . . . who
already have little opportunity to get to know each patient . . . and now will
face complex and constantly changing treatment options.)

Not everyone is on top of the speed or breadth of the genetic revolution.

We are way past the time when major genetic discoveries could be made . . .

In a little garden, fertilizing peas.

To cope with and organize the masses of data emerging from every lab . . .

It helps to have supercomputers . . .

Or massive parallel processing . . .

Machines that are concentrated in very few places.

Of the world's 500 largest computers. . . .*

255 are in the United States . . .

36 in the United Kingdom . . .

34 in Germany . . .

34 in Japan . . .

16 in France.

But European countries have especially strong anti-genetics lobbies . . .

Until recently, India and China had few resources, capital, and little respect for entrepreneurs . . .

So many great minds ended up flocking . . .

To the centers leading the life-science revolution . . .

Maryland . . .

Boston/Cambridge . . .

San Francisco/San Diego.

(And the local chambers of commerce and real-estate agents are thankful . . . Boston and San Francisco now suffer the most expensive office rents in the United States.)

(P.S. Singapore has launched a massive multibillion-dollar life-science initiative.)

*From www.top500.org.

As the digital and genetic revolutions converge . . .

So will their languages . . .

A, T, C, Gs . . .

And 1s and 0s . . .

Are interchangeable . . .

And may lead to . . .

Ever more powerful data processors . . .

Some of which . . .

Could be based on

DNA.

VIII

THE MOST POWERFUL INFORMATION SYSTEM

["DON'T EXPLAIN
WHY IT CAN'T
BE DONE.
DISCOVER HOW IT
CAN BE DONE."]

MO TAO

(404–319 B.C.)

In 1990, one of the world's best chess players, Garry Kasparov, faced off against Deep Thought, an IBM computer . . .

And beat it mercilessly.
In 1996, there was a rematch against IBM's Deep Blue . . .
At the time it seemed impossible that a machine could beat . . .
One of the great human minds . . .
In a game as subtle and complex as chess . . .
Except that the computer could study millions of positions on the board . . .
Every second.

Kasparov cried after he lost the first match . . .
But he rallied . . .
And went on to win three remaining matches.

Big Blue did not cry . . .
Computers get better at chess every day . . .
And people rarely do . . .
So at the 1997 rematch, Kasparov lost.

And today . . .
There is no chess player that can hope to beat a machine . . .
That can analyze more than fifty billion board positions every three minutes . . .

Soon no one will be able to beat a PC.

The machines required to decode the human genome had to be at least three orders of magnitude more powerful than IBM's Deep Blue . . .

So IBM's next-generation machine was called . . .

Blue Gene...[1]

(And may be able to carry out one quadrillion operations . . . every second.)

And at IBM's headquarters, many believe their company will be a life-science company . . .

Within a few years.

The ramp-up in computer power implies that . . .

By 2010, a computer should have the same processing capacity as a human brain.

And, if you are a teenager, by the time you have a few grandchildren (say, 2048) . . .

These kids may carry around the equivalent of $1,000 PCs . . .

With a processing capacity equal to all the human brains in the United States.

(These trends have led to a very public debate between two great minds, MIT's Ray Kurzweil, author of *The Age of Spiritual Machines: When Computers Exceed Human Intelligence,* and the chief technology officer at Sun Microsystems, Bill Joy, who wrote a very pessimistic essay for *Wired* titled "Why the Future Does Not Need Us." If these trends interest you, these are great pieces).[2]

Just as the Industrial Revolution allowed some people to multiply their physical capacity a hundred- or a thousandfold . . .

(And created large gaps between those who had machines and the education to use them and those who did not.)

The machines and technology coming out of the digital and genetic revolutions may allow people to leverage their mental capacity a thousand . . .

A million . . .

Or a trillionfold.

Biology is now driven by applied math . . . statistics . . . computer science . . . robotics . . .

The world's best programmers are increasingly gravitating toward biology . . .

You will be hearing a lot about two new fields in the coming years . . .

Bioinformatics and Biocomputing.

You rarely see bioinformaticians . . .

They are too valuable to companies and universities.

Things are moving too fast . . .

And they are too passionate about what they do . . .

To spend a lot of time giving speeches and interviews.

But if you go into the bowels of Harvard Medical School . . .

And are able to find the genetics department inside the Warren Alpert Building . . .

(A significant test of intelligence in and of itself . . . Start by finding the staircase inspired by the double helix . . . and go past the bathrooms marked XX and XY . . .)

There you can find a small den where George Church hangs out, surrounded by computers.

He seems like a large teddy bear . . .

Often hiding his extraordinary brain behind a quiet grin, a beard, and a self-effacing manner.

This is ground zero for a wonderful commune of engineers, physicists, molecular biologists, and physicians . . .[3]

And some of the world's smartest graduate students . . .

Who are trying to make sense of the thousands of terabytes of data that come out of gene labs yearly . . .

A task equivalent to trying to sort and use millions of new encyclopedias . . . every year.[4]

You can't build enough "wet" labs (labs full of beakers, cells, chemicals, refrigerators) to process and investigate all the opportunities this scale of data generates.

The only way for Church & Co. to succeed . . .

Is to force biology to divide . . .

Into **theoretical and applied** disciplines.

Which is why he is one of the founders of *bioinformatics* . . .

A new discipline that attempts to predict what biologists will find . . .

When they carry out wet-lab experiments in a few months, years, or decades.

In a sense, this mirrored Craig Venter's efforts at The Institute for Genomic Research and Celera.

These were information centers . . . not traditional labs . . .

And a few smart people are going to be able to do . . .

A lot of biology . . .

Very quickly.

(The human genome was privately mapped in a single lab . . . albeit one six football fields long . . . that ran 24 x 7 . . . but employed only fifty people running sequencing machines.)

Few Wall Street analysts have gotten it yet . . .

Because they still look at genomics companies as they do drug companies . . .

Expecting a long and costly development process.

But these companies sell bio information . . . not molecules.

They are like Lexis-Nexis, like Bloomberg . . . not like Merck.

Lexis-Nexis takes masses of publicly available data like court cases and newspaper and journal articles, and makes it easy to search the data given any of a myriad of variables.

Merck does a lot of proprietary research, isolates one molecule in ten thousand, and builds a nineteen-year patent moat around this new drug.

THE HEART OF CELERA . . .
WAS THE WORLD'S LARGEST PRIVATE SUPERCOMPUTER . . .
FED 24 HOURS A DAY . . . BY SEQUENCING
ROBOTS . . . AND CREATED-PROGRAMMED-CONTROLLED . . .
BY A DOZEN GREAT MINDS.

Celera's computers could amass 1.3 teraflops of data. Now companies hold a slightly obscene-sounding "petabyte" of storage space . . . A petabyte is 1,000 trillion bits (1s and 0s) . . . Of course, after petabytes come exabytes, zettabytes, and yottabytes.

(What is it with these guys—did they use Dr. Seuss as a consultant?)

According to a U.C. Berkeley study, the entire world's print and electronic media was producing about 1.5 exabytes worth of data per year (i.e., 500 billion U.S. photocopies, 610 billion global e-mails, 7.5 quadrillion minutes of phone conversations, etc.). All words spoken by all human beings throughout history could be stored with around 5 exabytes.[5]

Celera had a few competitors . . .
Among them the publicly funded International Sequencing Consortium, which includes:

ALBERT EINSTEIN COLLEGE OF MEDICINE ✳ Baylor College of Medicine ✳ BEIJING HUMAN GENOME CENTER ✳ Institute of Genetics ✳ CHINESE ACADEMY OF SCIENCES ✳ Center for Genetics in Medicine (Perkin Elmer/Washington University) ✳ GESELLSCHAFT FÜR BIOTECHNOLOGISCHE FORSCHUNG MBH ✳ Genoscope ✳ GENOME THERAPEUTICS CORP. ✳ Institute of Molecular Biotechnology ✳ KAROLINSKA INSTITUTET ✳ Center for Molecular Medicine ✳ LITA ANNENBERG HAZEN GENOME CENTER ✳ Max Planck Institute for Molecular Genetics ✳ JOINT GENOME INSTITUTE ✳ JAPAN SCIENCE AND TECHNOLOGY CORP. ✳ RIKEN Genome Sciences Center ✳ THE SANGER CENTRE ✳ Stanford DNA Sequencing & Technology Center ✳ STANFORD HUMAN GENOME CENTER ✳ University of Utah Genome Center ✳ UNIVERSITY OF OKLAHOMA ADVANCED CENTER FOR GENOME TECHNOLOGY ✳ University of Texas Southwestern Medical Center ✳ University of Washington Multimegabase Sequencing Center ✳ WHITEHEAD INSTITUTE FOR BIOMEDICAL RESEARCH ✳ and the Washington University Genome Sequencing Center . . .

Pre-Celera, their target was to finish the genome by 2005.

After Celera announced its intention to sequence the genome . . .

The public target was shifted to 2003 . . .

Then a rough draft by 2000.

But Celera had better ideas . . .

Smart people who were willing to take risks . . .

So even though the competition was far larger . . .

Spent five times as much . . .

Had a ten-year head start . . .

Celera still won.

When the "rough draft" of the genome was announced on June 26, 2000 . . .

It was called a "tie" by politicians . . .

It wasn't. Here is what the numbers looked like at the time.

PRIVATE VS. PUBLIC SEQUENCING EFFORTS

	CELERA	NIH/HGP
SEQUENCING FINISHED	April 2000	85% by June 2000
BASIC ASSEMBLY FINISHED	June 2000	24% done in June 2000
DNA VARIANTS FOUND (SNPs [SINGLE-NUCLEOTIDE POLYMORPHISMS] JUNE 2000)	6,000,000	300,000
COST	$250 million	$3 billion ($300 million spent January 1999 to June 2000)

(But if you cannot win then . . . use political leverage. In what will one day be seen as a dark chapter of public research, the NIH's Human Genome Research Institute claimed that its consortium had sequenced a draft of the full genome in June 2000, even though its own data showed it far behind. They also attempted to fund Celera competitors and tacitly discouraged government grantees from cooperating with Celera. . . .)

(OK, OK . . . I know that it's Washington and that Francis Collins, NHGRI director, is a political appointee, but still.)

In the end, the public consortium was able to publish a rough draft of the full genome the same week as Celera (February 12, 2001).

An unlikely trio pulled Collins' castanets out of the fire.

James Kent, a U.C. Santa Cruz graduate student who wrote a computer assembly program that the public consortium had not realized it needed . . .

> Eric Lander, head of MIT's Whitehead Institute . . .
> One of the great minds running around this world . . .
> An ex-business school professor who reinvented himself
> as a biologist . . .
> Founder of Millennium Pharmaceuticals . . .
> Great speaker . . .
> Lead author on the public human genome paper . . .
> Who unfortunately felt it necessary to do everything
> in his power . . .
> To block the simultaneous publication of the
> Venter-Celera paper . . .
> (While arguing academic freedom, of course.)

Sir John Sulston . . .
A priestlike figure . . .
Who rails against private control of genomes . . . ("It is a criminal act . . .")
And who ran Britain's Sanger Centre . . .
Convincing his trustees to spend tens of millions . . .
To make sure the genome remained in the public domain.

> There were of course thousands of other actors in this quest to reverse-engineer the human body.

> And as historians and biographers delve through the ups and downs of the quest . . .

> They find there was enough anger, pettiness, ego, passion, and greed on all sides to keep Hollywood busy for years.

> There was also great science, will, and vision . . .

> And despite the plodding and vengeful Dr. Collins . . .

> Mankind was able to begin studying its complete genetic sequence years before the government's initial estimates.

Collins is not a lone example of those trampled by the speed of change.

THE RULES ARE DIFFERENT IN A

KNOWLEDGE ECONOMY . . .

IT'S A SCARY TIME FOR THE ESTABLISHMENT.

**Countries, regions, governments, and companies
that assume they are . . .
And will remain . . .
Dominant . . .
Soon lose their competitive edge.**

(Particularly those whose leadership ignores or disparages emerging technologies . . . Remember those old saws: The sun never sets on the British Empire . . . *Vive La France!* . . . All roads lead to Rome . . . China, the Middle Kingdom.)

Which is one of the reasons bioinformatics is so important . . .

And why you should pay attention.

What we are seeing is just the beginning of the digital-genomics convergence.

When you think of a DNA molecule and its ability to . . .

Carry our complete life code within each of our cells . . .

Accurately copy the code . . .

Billions of times per day . . .

Read and execute life's functions . . .

Transmit this information across generations . . .

It becomes clear that . . .

**The world's most powerful and compact coding and
information-processing system . . . is a genome.**

So it makes sense to think about trying to design computers merging silicon and DNA . . .

(after all, a cell is far smaller than a CD but carries orders of magnitude more information . . . And someday, self-assembled nano molecules might lead to handheld supercomputers.)[6]

Which is one reason why Jim Clark, founder of Silicon Graphics, Netscape, and Healtheon . . .[7]

Gave $150 million, the largest individual gift in Stanford's history . . .

To launch biosciences and bioengineering.

IF YOU CAN GROW COMPUTERS ORGANICALLY . . . IT WILL BE MUCH EASIER FOR THEM TO FIX THEMSELVES . . . TOLERATE DEFECTS . . . AND PROCESS VAST AMOUNTS OF INFORMATION.

Today . . .
If one of the logic circuits in your computer breaks down . . .
The machine or program crashes . . .
(And you say *!#&*%!!! . . .)
But if PCs acted like the Internet . . .
And could bypass disabled nodes . . .
They would be far more reliable.

One way to accomplish this . . .

May be to blend silicon chips with DNA . . .

So memory chips could grow organically . . .

And pack a lot more
processing power
Into much
Smaller spaces.[8]

If all this sounds unlikely to you . . .

Consider the principles guiding IBM's computer architecture.

The acronym is SMASH . . .

"Simple, Many, And Self-Healing."

Computers are increasingly designed to act as collections of individual cells . . .

That work in parallel to solve problems . . .

And can fix themselves.

> Programmers are stringing together computers into
> "Beowulf clusters" . . .
> Where hundreds or thousands of computers are wired together
> using Linux software . . .
> And mutate various programs in an attempt to accomplish
> a task.
> After various cycles, only the "fittest" programs survive.

> **(If you want, you can try to apply this Darwinian solution in your home or office today.)**[9]

But before we get to lifelike computers . . .

Various things have to happen.

We have to be able to read the code of life . . .

(What Venter and the Human Genome Project are doing today.)

We have to understand how life code becomes life processes.

In other words, how tens of thousands of genes code the million-plus proteins that regulate our lives daily . . .

Companies are spending billions . . .

To increase their computing power by orders of magnitude . . .

An attempt to code all proteins and their interactions with one another.

Which is one of the reasons you hear a lot not just about genomics . . .

But also about proteomics.

(Proteomics is orders of magnitude more complex than sequencing the human genome, because instead of dealing with tens of thousands of genes, you're dealing with millions of different proteins and perhaps millions of distinct expressions of these proteins.)

> The first protein structure was unraveled in 1957 . . .
> By 1990, only 589 protein structures were publicly available.
> And it cost hundreds of thousands of dollars to unravel each protein . . .
> A process that took three to five man-years.
> But things are getting a lot faster.
> By mid-2000, there were 12,777 proteins published.[10]
> Argonne National Lab had characterized a 3-D protein crystal structure in 6½ hours.
> In January 2001 a company claimed it had 115,693 distinct proteins.
> Some now want to sequence thousands of proteins an hour.

(Many companies are focusing on proteomics . . . For example, AxCell Biosciences, Ciphergen Biosystems, Incyte Genomics, Genebio, Genomic Solutions, Proteome, Oxford GlycoSciences, Large Scale Proteomics, Life Technologies, Myriad Genetics, Protana, most major pharmaceuticals.)

*One way to understand how genes code and what
they do is through gene chips . . .*

*Chips are already far more than simply memory
units of computers.*

Affymetrix bought an old National Semiconductor manufacturing
facility . . .

And polished the silicon used in computer chips.

But instead of placing transistors on the chip . . .

It placed hybrid bits of DNA . . .

Which come together like a zipper with biological samples from your
body.

(If you happen to have the health conditions that the chip tests for . . .)

If the particular place where your DNA and the sample placed on the
chip match . . .

It lights up under a laser.

These chips can determine whether an individual has any one . . .

Of 60,000 genetic conditions.

Within five years . . .

An Affymetrix chip may include markers for all human genes . . .

And test for hundreds of thousands of genetic variants.

("Yippee, a new business opportunity," said other companies as they
also dove into this pool: Hewlett-Packard [Agilent], Motorola, Corning,
and Hitachi, as well as a myriad of start-ups with names like BioTrove
and GeneXP.)

The reason these changes are occurring so quickly is because . . .

It is not a single technology or field that is driving change.

Chips are the result of convergence and synergy among various technologies . . .

Like inkjet printers . . .

Which can accurately place millionths of a liter of liquid.

Automated lasers . . .

Which can read and store a lot of information very quickly.

Electron microscopes and instruments . . .

Which can move individual atoms.

Computers . . .

Which can analyze and correlate millions of variables.

Polishing pads and liquids . . .

Which allow you to build seven stories' worth of electronics on a single chip.

Some gene chips can heat, cool, distill, precipitate . . .

In other words, they are chemistry labs . . .

The size of a quarter.

If these technologies seem overwhelming to you . . . you are in good company.
In the 1870s, a bishop of the Church of the United Brethren in Christ was outraged.
By a college president's suggestion that someday . . .
"Men will fly through the air like birds."
The bishop angrily responded, "Heresy . . . blasphemy . . . flight is reserved for angels."
Then Milton Wright went home to care for his two children . . .
Wilbur and Orville . . .
Bicycle mechanics . . .
Who used "Pride of the West" ladies' underwear cloth . . .
To cover the wings . . .
Of the world's first airplane.[11]

Gene chips will lead to personalized medicine . . .

You will be able to test whether one medicine or another works better for you . . .
Before you take it.

You cannot do that today . . .

Which is why when you buy a drug . . .

And open the box . . .

The first thing you get . . .

Is a long list of the ways in which . . .

The prescription you paid for . . .

May hurt you.[12]

(I.e., this medicine may cause bleeding,

cramps, dry mouth, impotence, insomnia, irritability,

nausea, and other fun stuff.)

Thalidomide is one example of highly personalized medicine . . .

Few images are as devastating as the pictures of deformed children published in *Life* . . .

Caused by prescribing a "safe sedative" to pregnant women.

Ten thousand children worldwide were affected . . .

The drug was *verboten* globally for forty years . . .

But now this compound may help treat . . .

Leprosy, cancer, AIDS, rheumatoid arthritis, inflammatory bowel disease, tuberculosis.

This is one example of a medicine that is beneficial to many . . .

But highly toxic to some.

Many drug launches have died in Phase III trials . . .

After millions of dollars are spent.

By this point you know the drug helps some people . . .

But maybe only a few . . .

Or maybe it really hurts some folks . . .

So you kill the drug . . .

Because you cannot personalize treatment.

This is about to change . . .

And smart companies are going to go back to their files and labs . . .

To revive a lot of compounds . . .

Very soon.

Gene chips will also tell you if you have a higher predisposition . . .
To certain diseases . . .

And what variants of the diseases you may get.

DOCTORS
WILL FOCUS MUCH
MORE ON PREVENTION . . .

AND LESS ON TREATMENT.

In a sense, medicine will follow dentistry.

Our grandparents went to the dentist to get teeth pulled . . .

We went to dentists to get cavities filled.

Our kids go to dental hygienists . . .

And get their teeth cleaned.

Rarely do they see the dentist . . .

(Until they get braces.)

We protect our teeth . . .

By brushing, flossing, drinking fluoridated water.

We will protect our bodies . . .

By eating neutraceutical foods . . .

Using soaps and cosmetics . . .

Taking daily pills . . .

That are targeted to our specific genetic conditions.

Bill Haseltine believes we will see a seismic shift in how
we cure people.

Today, for every dollar we spend on medicines . . .

We also spend nine on treatments and interventions.

Haseltine thinks we will soon spend at least an equal amount . . .

On prevention . . . just as we do with dentists.

We often forget just how fast the way we live changes.

A worker in 1897 paid the equivalent of . . .[13]

$	1	Per pencil
	62	Webster's dictionary
	67	Scissors
	1,202	Phone
	2,222	Bicycle

Our grandparents rarely flew to other countries . . .

And calling London from New York . . .

Used to cost the equivalent of $200 . . .

For three minutes.

(But at least they could make a call . . . When Abraham Lincoln was murdered, it took fifteen days for the news to reach London.)

Grandpa's lifestyle seems very primitive to us . . .
Just as **today's medicine** . . .
Will seem like voodoo . . .
To our grandkids.

When M.D.s look at a cancer biopsy . . .

The first thing they search for is . . . positive or
negative . . . malignant . . . not malignant.

Often, the second is . . . which technician did the analysis . . .

Because results are highly subjective.

Tumors that look the same respond very differently to
the same treatment . . .

And hurt people in different ways . . . at different rates . . .

So harried doctors have to learn whom they can trust . . .

(And whom they have to double-check.)[14]

(Someday you will be asked: "Grampy . . . are you pulling my leg when you tell me
they used to poison people and cut their body parts off to try to cure cancer?" . . .
Think I am exaggerating how barbaric we will seem? Consider acceptable treatments for schizophrenia during the mid-twentieth century . . . lobotomies, elec-
troshock, and infecting the patient with malaria.)

Most medicine still treats mostly symptoms . . .

So even relatively simple diseases like ulcers . . .

Keep reminding us of how ignorant we still are.

In the 1950s hospitals treated ulcers with antibiotics . . .

It worked.

Then M.D.s decided that the cause of this distress was not disease . . .

But environmental pressures like angst, overwork, fear . . .

They made a mistake because they focused on a symptom . . .

Increased stomach acid.

So they prescribed a series of best-selling drugs . . .

Which make you feel better but ignore the problem . . .

Because ulcers are caused by a bacterium, *Helicobacter pylori* . . .

(a genome sequenced by The Institute for Genomic Research [TIGR]) . . .

And it turns out that when you reduce the acid in your stomach . . .

The bacterium that causes the disease is also happier . . . and keeps reproducing . . .[15]

Genomics, bioinformatics, and biocomputing are merging . . .

(In gene start-ups like DoubleTwist, 80 percent of the 110 Ph.D.s . . . were computer scientists.)

And they are beginning to explode a lot of myths . . .

Medicine is going to change . . . . fast.

P.S.

Just as Celera started to get comfortable with its wonderful new computer center . . .

IBM announced that a private company with thirty-eight employees . . .

NuTec Sciences . . .

Had just purchased a machine capable of doing 7.5 trillion calculations per second . . .

And expected to operate the world's largest commercial computer center . . .

To help academics and drug companies . . .

Process the waterfalls of data coming out of the Human Genome Project.

P.P.S.

In January 2001 . . . Celera, the U.S. Department of Energy, and Compaq announced a partnership to create a computer that will process 100 trillion operations per second.

IX

NANO WORLD

.

We can already build pretty small things . . .

Japanese engineers have produced a car . . .

That looks and operates like any other car . . .

Except that it is the size of a grain of rice.

Fun . . .

But trivial compared to what
is coming.

The first computer chips were marvels . . .

They integrated hundreds of circuits . . .

On a small piece of silicon.

But building faster, more powerful computers required . . .

Packing more and more components in a smaller space . . .

And we have gotten very good at this.

Some chips . . .

Built this year . . .

Should contain close to a billion components.

When manufacturers polish the surface of chips today . . .

The error tolerance is 0.01 microns . . .

(A micron is a millionth of a meter . . .)

Or about one hundred-thousandth the width of one of your hairs.

But these manufacturing standards will soon seem crude . . .

Because various companies and labs are now experimenting with . . .

Products built on a nano scale.

When you build on a nano scale. . . .
Things are measured in billionths of a meter.
In other words . . .

You can write out "IBM" using individual atoms . . .

Something that was extraordinary when it was accomplished . . .

In 1989.

By 2000, Northwestern University's Chad Merkin programmed
an atomic-force microscope to write paragraphs on a nano scale
using a computer keyboard.

To put this in perspective . . .

You could write out the whole *Encyclopedia Britannica* . . .[1]

On the head of a pin . . .

Or print most of the world's books on six square yards of
silicon.[2]

But even this is relatively trivial compared to the ability to
build . . .

A rotor on a molecular scale . . . as IBM Zurich did.[3]

IF YOU EVENTUALLY BUILD A MOLECULAR MOTOR . . .
YOU CAN POWER . . . MACHINES THAT CAN FLOAT
(LITERALLY) ON A SPECK OF DUST.

Such machines may be possible, thanks to three cantankerous
scientists . . .
Robert Curl Jr., Harold Kroto, and Richard Smalley . . .
Who fought each other . . .
And the world . . .
To discover and build . . .
New molecules . . .
Out of the world's most studied element . . .
Carbon.[4]

Before 1985, no one knew . . .
That you could lattice carbon . . .

To build geodesic molecules (which look like soccer balls) . . .

On a nano scale.

The structure of these
compounds resembles the domes
that a famous architect . . .

Named R. Buckminster Fuller...

Used to design . . .

So they are called fullerenes . . .

And may be the key to building nano scale . . .

Medicine transporters . . .

Super-strong tubes . . .

Transistors.[5]

In February 2000, IBM scientists announced that they
were starting to think about designing a computer on a
molecular scale . . .

A computer that could float through the air.

A seemingly crazy idea until you realize that ENIAC . . .

The first all-electronic digital computer (1946) . . .

Weighed thirty tons and was 100 feet long . . .

But was far less powerful than the chip in your PC, which
can fit on your fingertip.

(ENIAC could do fourteen 10-digit calculations per second; IBM's supercomputer, Blue Gene, may do close to a quadrillion.)

If you bring computers together with three-dimensional printers . . .

You can design and build robots . . .

Without any additional human input.

And these robots can evolve . . .

Build other robots . . .[6]

And, eventually, perhaps alternate life forms.

In November 2000, Cornell University scientists built "nanosubs."

These devices are partly organic and partly inorganic . . .

They run on the same stuff that powers your cells (ATP) but incorporate nickel propellers.

They can run for two and a half hours . . .

And are about as large as a virus.

Someday, "nanonurses" may be able to seek specific diseased cells within your body . . .

And deliver molecular doses of medicine.[7]

(You can now routinely look at, move, and even feel individual atoms and molecules . . . something that used to be described as "playing blindfolded billiards during an earthquake.")[8]

Lest this talk of new life forms seems abstract or farfetched . . .
Reflect for a minute on what genomics has already un-taught
us about biology.

Methanococcus jannaschii's genome led to a crisis among those
who publish biology textbooks.

These obscure bacteria were first found in Yellowstone's boiling
pools, where the liquids are as corrosive as battery acid . . .

Then they were dragged up from the deepest parts of the
sea by a small sub called *Alvin.*[9]

They live happily next to underwater volcanoes, where
continental plates come apart.

In complete darkness . . . swimming in boiling water . . . under
an atmospheric pressure 245 times that of land . . .

They love carbon dioxide . . . are poisoned by oxygen . . . can
drift through close-to-freezing water to another volcanic vent.

These organisms were different from any life form encountered, to that
point, on the planet.

Fifty-six percent of their 1,738 genes were unlike those of any other
species.

But it turns out they have been around for over three billion years . . .

Since the time the earth was hot, covered with volcanoes and poisonous
clouds . . .

So they were called *Archaea* . . . ancient ones.

And their discovery forced biologists to admit that there was a third
branch on the tree of life . . . one that could breathe iron instead of
oxygen.

(The two traditional branches were *prokarya* [bacteria] and *eukarya* [fungi, proto-
zoa, algae, plants, and animals]. This division became unsustainable after Venter
and Claire Fraser published "Complete Genome Sequence of the Methanogenic
Archaeon, *Methanococcus jannaschii.*")[10]

After biologists realized that life did not have to follow their "rules" . . .

They opened their eyes and minds . . .

And found *Archaea* everywhere . . .

In salt flats, inside the core of the Earth, inside volcanic pools with a pH level equivalent to sulfuric acid, in rice paddies, in wine by-products.[11]

To make a long story short . . .

A life form discovered a few years ago may account for . . .

One-fifth of the biomass on Planet Earth.

> As we find more and more creatures that can thrive in extreme environments . . .
>
> The likelihood of life on other planets increases.
>
> Think for a second of what it means to find a funky bacterium . . .
>
> Called *Deinococcus radiodurans* swimming happily inside your neighborhood nuclear reactor . . .
>
> Exposed to doses of radiation that would immediately kill you.[12]
>
> Or what it means to find a 250-million-year-old bacterial spore . . .
>
> That was buried 1,500 feet below the earth's surface in a Carlsbad, N.M., salt dome . . .
>
> That grew into a bacterium similar to a bacillus when brought to the surface and fed.[13]

Meanwhile, over at U.C. Santa Cruz . . . biologists replicated the vacuum of space (no air, extreme cold, high radiation) . . . and were still able to grow primitive cell walls.[14]

(And given that astronomers have now discovered extra solar planets . . . and find more every month . . . Still believe we are unlikely to discover life on any other planet in the universe? That we are the be-all and end-all? . . . No worries—some people still believe the sun revolves around the Earth or that evolution is a false doctrine.)

Genomics is teaching us what life really is . . . on a nano scale.

After sequencing the organism with the smallest amount of genetic material, *Mycoplasma genitalium* (517 genes) . . .

Craig Venter and Claire Fraser went back to their lab and systematically began extracting genes one at a time . . .

And observing whether the creatures died . . .

An experiment that showed exactly which genes (instructions, if you will) are essential to stay alive.

After eliminating 180 genes . . .

They were within a few genes of knowing what it takes to live.

But instead of spending one more day in the lab . . .

They stopped the experiment . . .

And asked a panel of bioethicists—including Catholic, Jewish, and Protestant theologians—whether they should continue . . .

Because the answer could be the basis for building life in a test tube.

(The panel said, "Go ahead.")[15]

I don't want to convey that biology is where all the action is . . .
Or that these four revolutions (genomics, proteomics, biocomputing, nanotech) . . .
Are all that matter.

What is extraordinary is how widespread and powerful . . .

The technology base of the knowledge economy . . .

Is getting and how discoveries in one area . . .

Lead to breakthroughs in other fields . . .

Quickly.[16]

Even art is changing.
It is hard for an artist to come up with anything that will really shock people anymore . . .

But in 2000, Eduardo Kac did . . .

By exhibiting a fuzzy white rabbit, Alba . . .

Whose DNA includes genes from a phosphorescent jellyfish . . .

So when exposed to black light, every cell in the animal's body glows bright green.

IS BECOMING AN ARTISTIC MEDIUM.[17]

(Using phosphorescence for nonscience purposes is an old pastime: 2,000 years ago Romans discovered a clam that made one's lips glow when eaten, so they would make the room dark and cavort.)

Each field reinforces and accelerates discoveries in each of its neighbors . . .

CONVERGENCE AND ACCELERATION . . .

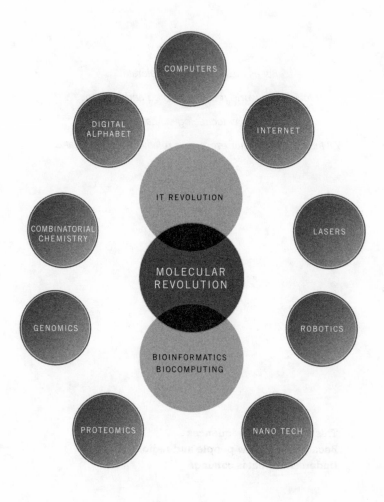

But as these changes accelerate . . .

Much of the world remains isolated and ignorant . . .

About technologies that . . .

Will fundamentally alter everyone's relative economic status . . .

And life expectancy.

This will have consequences . . .
Because only a few people and regions . . .
Understand what is coming.

REVOLUTION . . . IN A FEW ZIP CODES

["THE WORLD IS
DIVIDED INTO A
GOLDEN BILLION . . .
AND EVERYONE
ELSE."]

VLADIMIR PUTIN
PRESIDENT OF RUSSIA

As new disciplines and technologies emerge . . .

You don't win the game by just producing gobs of knowledge . . .

You also have to protect it . . .

And apply it.

Which is why patents are a good barometer . . .

Of creativity . . .

Tenacity . . .

Ability to articulate an idea . . .

And capacity to build knowledge.

Patents are a good window . . .

(Although not the only window) . . .

On who might triumph . . .

And who might lose . . .

Over the course of the next two decades.

Not all patents are good . . .

Or valuable . . .

But being unable to generate patents . . .

IS

VERY

BAD.

TO COMPETE GLOBALLY . . . ONE HAS TO PATENT GLOBALLY . . .
PARTICULARLY IN THE UNITED STATES AND EUROPE.

In 1985 . . .

The U.S. Patent Office granted Argentines 12 patents . . .

Venezuelans got 15 . . .

Brazilians 30 . . .

And Mexicans 35.

South Korea got 50 patents.

In 2003 . . .

The same office granted Venezuelans 20 patents . . .

Argentines 70 . . .

Mexicans 92 . . .

And Brazilians 180 . . .

South Koreans received 4,132 patents.[1]

In other words . . .

During 2003, 11,592 Koreans generated enough knowledge to obtain one U.S. patent . . .

It took 1,140,865 Mexicans . . .

Or 1,022,784 Brazilians . . .

To accomplish the same task.

The average real wage of Koreans multiplied ninefold between 1960 and 1990.

The real minimum wage in Mexico was about the same in 2000 as in 1960.[2] By 2000, it had fallen further.

As education and knowledge accumulated, this trend became self-reinforcing; between 1990 and 1998 Korea's real economic-growth rate was eight times larger than Mexico's.

MINIMUM WAGE IN MEXICO
(1997 pesos)

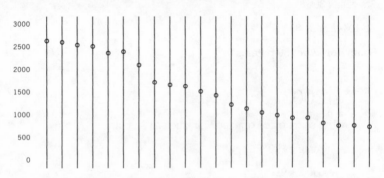

68% of Mexico's population earns the equivalent of two minimum wages or less, which tells you that most people earn less than they did two decades ago.

If you cannot create knowledge and patents . . .

It is impossible to launch new high-tech companies.

As the old companies become uncompetitive . . .

There is nothing to replace them . . .

Which is the main reason why . . .

The number of companies listed on the Buenos Aires
stock exchange . . .

> Dropped from 149 in 1995 . . .

> To 125 in 2000.

At the end of the twentieth century . . .

there had been no IPOs . . .

In major Latin American exchanges . . .

For close to three years.

(And between 1990 and 1996, the number of new books
published in Mexico fell from 21,500 to 11,762 . . . Meanwhile,
during 2000, over four hundred companies were IPOed
in the United States.)

There was little new knowledge to sell . . .

> And

> Little

> Economic

> Growth.

(Not surprisingly, 49 percent of U.S. households own some
stocks, while in Mexico less than 0.5 percent do.)

What matters in a modern economy is knowledge.

It is what you live off . . .

It is what powers growth . . .

And from patents in the last twenty years . . .

It is not that hard to predict who gets rich . . .

And who gets poor.

PEOPLE NEEDED TO PRODUCE A SINGLE U.S. PATENT

COUNTRY	PEOPLE PER PATENT (2003)
United States	3,308
Japan	3,592
Taiwan	4,266
Switzerland	5,619
Germany	7,213
Canada	9,232
Singapore	9,953
South Korea	12,148
France	15,446
United Kingdom	16,358
Australia	22,090
Spain	133,013
Argentina	583,683
Mexico	1,217,750
Venezuela	1,351,263
Brazil	1,358,431
India	3,121,405
Indonesia	23,852,667

The lack of ability to generate knowledge . . . reinforces itself . . .

The best researchers . . .

    The most entrepreneurial folk . . .

        Want to be where the action is . . .

           Where they can find . . .

               Information . . .

                  Challenges . . .

                     Respect . . .

                        People . . .

                           Funds.

So **they move**.

And when they do . . .

The overall wealth of a region . . .

Or of a country . . .

Also increases or decreases.

FROM 1970 TO 1995,

THE WORLD'S RICHEST COUNTRIES

GREW 1.9 PERCENT PER YEAR ON AVERAGE . . .

MIDDLE-INCOME COUNTRIES GREW 0.7 PERCENT . . .

THE POOREST THIRD DID NOT GROW AT ALL.[3]

In 1997, the Brazilian patent and trademark office approved 70 percent fewer requests for intellectual-property protection . . . than those approved in 1982.[4]

In part this occurred . . .

Because Brazil is now part of a global system . . .

That respects intellectual property established in other countries . . .

So many companies do not bother to file a separate Brazilian patent.

But it is also true that . . .

Brazil is generating less breakthrough knowledge . . .

And Brazil is less relevant in the global-knowledge market.

Brazil is not alone.

In 1997, the Mexican patent office received 10,531 patent requests. . . .

Only 4 percent of the requests came from Mexican nationals . . .

And the overall trend is not good.

In **1988,** Mexican citizens presented **652 patent requests . . .**

Within their own country . . .

In **1997,** there were **420 requests.**[5]

In 2003, **468** and still less than **4% of the total.**

When one looks at who is producing knowledge in Mexico, there is not a single Mexican company or organization among the top ten patent grantees.

WHO PATENTS IN MEXICO? NOT MEXICANS . . .[6]

COMPANY	# OF PATENTS
Procter & Gamble	396
Kimberly-Clark	296
Bayer	246
Thomson Licensing	232
BASF	221
Astra Zeneca	120
QUALCOMM	106
Johnson & Johnson	100
Unilever	95
GE	94

The same trends hold abroad . . .

In **1968,** Mexicans applied for **177 U.S. utility patents** . . .

In **2003,** they received **92 patents** (0.032% of total patents).

Things do not have to be this way . . .

Given the right training, capital, and instruments . . .

Brazilian and Mexican scientists can be competitive.

São Paulo gives 1 percent of its tax revenue to a research institute, FAPESP . . .

And in 2000, FAPESP published the world's first plant-disease genome, *Xylella fastidiosa,* which destroys oranges, grapes, almonds, plums, peaches, oaks, and elms.

The discovery was not a fluke . . .

A few days later . . .

The institute published markers for 279,000 pieces of DNA (ESTs) . . .

Only the United States and Britain had published more.

> But this is only one institute . . .
> Within one state . . .
> Within a large country . . .
> It is an exception.

But what is happening to countries like Brazil and Mexico overall . . .

Is not exceptional.

**Most developing countries are not competitive in the global
knowledge economy . . .**

In 1998, a single company, IBM, obtained more U.S. patents . . .

Than the total granted to 139 countries.

(In 1999, IBM obtained even more U.S. patents, 2,756 . . . In 2003, it got
3,415 . . . That means that over the course of the 1990s, IBM created
enough knowledge to get more than 15,000 patents . . . If you have a lot of
time on your hands, you can browse them all at www.patents.ibm.com.)

IBM GENERATED MORE PATENTS ALONE . . .
THAN 139 COUNTRIES DID TOGETHER
(u.s. patents granted, 1998)

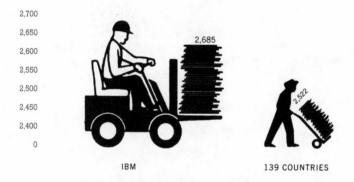

| | IBM | 139 COUNTRIES |

2,700
2,650
2,600
2,550
2,500
2,450
2,400
0

2,685

2,522

Seventeen countries generated 95 percent of U.S. patents in 2003 . . .

Most of the other 188 countries . . .

Are going to have trouble keeping up . . .

As technology and knowledge turbocharge economic growth.

The extraordinary tech revolution . . .

Is fed by a very few ZIP codes . . .

Generating new empires . . . (and new ghettoes).

It is not just companies that are accumulating global brainpower and creating unprecedented wealth.

Massachusetts Institute of Technology alumni and faculty have founded more than 4,000 companies . . .

Generating over $230 billion in yearly sales . . .

Which in terms of national economies . . .

Makes it twenty-third in the world.

Millennium Pharmaceuticals alone . . .

A midsize genomics company founded by an MIT professor . . .

Had 1,500 patents pending in 2000 . . .

And was worth more than everything produced . . .

In Lithuania plus Estonia over the course of a year.

Many of MIT's most successful alumni and faculty . . .

Are foreigners who chose to stay in the United States.

In a borderless world . . .

Those who do not educate . . .

And keep their citizens . . .

Will lose most intellectual wars.

(The United States has gotten lazy in this area . . . It prefers to import brains rather than generating them in its high schools . . . More on this later.)

The consequences of free trade . . .

In a technology-driven world . . .

Are different from having open borders . . .

In a commodity-driven world economy.

We expect countries and businesses to "compete freely" . . .

Even if the relative knowledge they generate . . .

                                        May be minuscule.

The United States spends around 2.6 percent of its yearly economic output on research and development.

Mexico spends about 0.3 percent of its GNP on R&D . . .

And the U.S. economy is "slightly" larger.

So the United States spends over $182,000,000,000 each year to improve its knowledge base . . .

And Mexico spends $1,400,000,000 . . .

A 130-fold difference.

The University of California system alone spends 21 percent more on R&D . . .

Than does the whole of Mexico . . .

And in the process generates six times as many U.S. patents.

(Not to mention companies . . . In 2004, Pfizer spent over $7.4 billion on R&D.)

AS A DEVELOPING COUNTRY . . .

YOU CAN LOWER INFLATION . . .

REDUCE CORRUPTION . . .

CUT YOUR BUDGET . . .

PRIVATIZE . . .

AND STILL NOT GET RICH . . .

Because you are not generating knowledge . . . just product . . .

(North America, Western Europe, and Japan generated 84 percent of all scientific papers published during 1995.)

Chile, often cited as the shining example of Latin American economic reform . . .
Carefully followed the recommendations of the most orthodox Ph.D.s in economics . . .
Nicknamed the "Chicago Boys" . . .
For a decade, its economy grew spectacularly.

But even Chile may be headed toward a crash . . .

Because it took the inefficiency out of the old economy . . .

But failed to build a new economy.

Two commodities, copper and cellulose, represent 40 percent of Chile's total exports . . .

Most of what Chile exports contains very little technology.

One can get a sense of how knowledge-intensive an economy is by dividing:

$$\frac{Value\text{-}Added\ Exports}{Commodity\ Exports}$$

If the resulting ratio is *greater than one, the country exports more knowledge-based* products than raw materials.

If it is *less than one,* the economy remains *vulnerable to commodity cycles.*

The knowledge-export ratio for Chile is even lower than the Latin American average.

HIGH-TECH VERSUS LOW-TECH EXPORTS[7]
(if index > 1, value-added exports exceed commodities)

YEAR	CHILE	ARGENTINA	BRAZIL	G-7 (CANADA, FRANCE, GERMANY, ITALY, JAPAN, U.K., U.S.)	ASIAN TIGERS (HONG KONG, SOUTH KOREA, SINGAPORE, TAIWAN)
1977	0.01	0.12	0.22	2.17	0.74
1995	0.01	0.07	0.23	1.67	1.80

There are extreme differences between what the world's most advanced countries (G-7) and the Asian tigers export . . .

And what Latin America, Africa, and the rest of Asia export.

If you do not export knowledge, you do not get rich.

You can see this clearly in the case of Mexico . . .

Which tripled its high-tech export ratio in less than two decades . . .

And achieved statistics comparable with those of the developed world or Asia's tigers.

TECHNOLOGY EXPORT INDEX (1995)

LATIN AMERICA	MEXICO	G-7	ASIAN TIGERS
0.5	1.62	1.67	1.8

But as exports increased, most people's real income remained the same . . .

Or fell.

Most cars, computers, and electronic exports from Mexico . . .
Are assembled in maquiladoras (in-bond factories) along the U.S. border . . .
And their national content and local value added are small . . .
Less than 3 percent of export value . . .
Because Mexico generates little proprietary knowledge.

(Maybe that is one reason why General Electric's CEO made more in 1999 than the combined wages of 15,000 maquiladora workers who assembled products for GE.)

So Mexicans, Brazilians, Argentines, Chileans, Africans, and Indians . . .

Keep restructuring their economies time and again . . .

But still tend to remain poor . . .

And face bleak prospects . . .

Because they generate and sell little new knowledge.

Meanwhile . . .

Within the United States and a few other enclaves . . .

Knowledge is cascading . . .

And so are riches.

(From 1900 to 1910 and from 1920 to 1930, none of the Nobel prizes in medicine went to Americans . . . During the 1990s, just one out of ten was awarded to non-Americans.)

The 1990s were characterized by low inflation . . .
But the number of U.S. millionaires increased . . .
From 1.3 million in 1989 . . . to more than 5 million in 1999 . . .
And if you want to count billionaires . . .
In 1982, there were 13 . . . by 2000, there were more than 250.

(But like in the movie *Gone with the Wind* . . . more than a few burned with the NASDAQ crash.)

Rarely do societies generate a lot of billionaires . . .

The previous U.S. crop (in real dollars) occurred during the Industrial Revolution.

It takes an economic discontinuity to accumulate this much wealth (legally).[8]

Today's discontinuity is the knowledge economy.

Just as occurred during the Industrial Revolution, the United States is starting to live through an explosion of knowledge and patents.[9]

U.S. PATENT ACTIVITY

YEAR	PATENTS GRANTED	Δ% INCREASE OVER 20 YEARS EARLIER
1840	458	—
1860	4,363	853
1880	12,926	196
1900	24,656	91
1920	37,057	50
1940	42,237	14
1960	47,169	12
1980	61,819	31
2000	182,223	195

(The first modern U.S. patent was granted July 31, 1790 and signed by George Washington . . . Three patents were issued that year: improvement of potash ash, a new way to make candles, and better flour-milling machines.)[10]

But before anyone gets too congratulatory and smug . . .
They should look at two trends . . .
Who patents . . . and where patents are generated.

OF THE TOP TEN COMPANIES
PATENTING IN THE UNITED STATES BETWEEN 1969 AND 2003 . . .
FOUR WERE AMERICAN . . . IBM, GE, KODAK, AND MOTOROLA . . .

THE OTHER SIX WERE . . .
CANON . . .
HITACHI . . .
TOSHIBA . . .
NEC (NIPPON ELECTRIC CORP.) . . .
MITSUBISHI . . .
MATSUSHITA . . .

Japan received almost 20 percent of all U.S. patents issued in 2003.

Meanwhile, the United States' largest trading partner . . .

Canada . . .

Got 2 percent . . .

So it should surprise no one . . .

That in a knowledge-driven global economy . . .

In 2000, the U.S. trade deficit with Japan was 70 percent higher . . .

Than it was with Canada.

Meanwhile, the Canadian dollar . . .

(Affectionately called "the loonie" because of the bird on the coin . . . and also perhaps because of the way it behaves in the currency markets . . .)

Continued to drop like a stone versus the U.S. dollar . . .

(From 1.04 per U.S. $ in 1974 to 0.65 in 2000 . . .)

Even though the Canadian economy was growing at a record pace through the 1990s . . .

And the country maintained a record trade surplus.

Because Canada's economy continues to depend primarily on forestry, agriculture, and mining . . .

And many of its smartest people leave . . . or simply list their companies on NASDAQ.

(No wonder New England toll collectors get testy when you mistakenly give them a Canadian coin.)

(But post-deficit, post-Bush II, the loonie flies and the dollar sinks.)

The United States' second-largest trading partner . . .

Mexico . . .

Got 0.1 percent of the 1999 U.S. patents . . .

And has continually devalued its currency over the past decades . . .

(One U.S. dollar equaled 12.5 pesos in 1975 . . . 257 in 1985 . . . 2,813 in 1990 . . . 6,450 in 1995 . . . 11,300 in August 2004.)

In a desperate effort to keep up exports to an ever more productive United States . . .

In other words, because Mexico cannot produce a lot of new knowledge . . .

It has to sell its labor at ever-lower rates . . .

Across an increasingly open border.

THIS IS NOT A RECIPE FOR LONG-TERM HEMISPHERIC STABILITY.

As the hemisphere falls further and further behind the United States . . .

In the knowledge economy . . .

It gets harder . . .

To reduce income disparity . . .

Defend open markets . . .

Promote democracy . . .

Control immigration . . .

Fight guerrillas . . .

Limit drugs.

Throughout Latin America, over the past two decades, the stereotypical mirrored-sunglasses military dictator went back to his hacienda . . . Democracy flourished . . . Markets opened . . .

And people got poorer.

There is little reason to believe—given recent crises in Argentina, Bolivia, Brazil, Colombia, Ecuador, Mexico, Paraguay, Peru, and Venezuela—that the hemisphere is more stable. Rather, a lot of people seem quite fed up with their governments, which leads to wild electoral swings between left and right.

Presidents still get ousted regularly . . . (Argentina's Alfonsín, Ecuador's Bucaram and Mahuad, Brazil's Collor, Peru's Fujimori, Paraguay's Cubas, Venezuela's Pérez . . . The same is true farther afield . . . Indonesia, Thailand, the Philippines.)

And the International Monetary Fund, the World Bank, and all those who declared that Latin America had turned a corner and was well on its way to fulfilling its potential . . .

Are flummoxed yet again.

Until Latin America and Canada pay more attention to science and technology and less attention to the current political soap opera . . . things are not going to get better.

The United States' neighbors are not benefiting or growing at the same rate as the U.S. is . . .

And, at some point, their interests and outlook may shift drastically . . .

Away from an economic model that leaves them ever further behind.

Because *if you tear down borders and fences . . .*

And bring coyotes and hens together . . .

The only way the hens survive . . .

Is if they are very smart hens . . .

And there are ever fewer such creatures running around Latin America, Quebec, and Africa.

Because more and more of them now live in Silicon Valley.

There are challenges within the United States as well.

Knowledge also concentrates regionally . . .

> **Five states . . .**
>
> **California, New York, Texas, Massachusetts, and Michigan . . .**
>
> **Generate 44 percent of all U.S. patents.**
>
> **So even inside the United States research was concentrated . . .**
>
> **Within a very few ZIP codes.**

(33 percent of all U.S. patents came from ten cities: San Jose, greater Boston, Chicago, Los Angeles, Minneapolis, Detroit, Philadelphia, New York, Rochester, San Francisco . . . 52 percent from twenty cities . . . And actually it was very specific research clusters within these cities . . . Perhaps that is why they call 02138, Harvard-Cambridge, the most opinionated ZIP code in the world?)

As the United States experienced unprecedented economic expansion and wealth creation . . .

When unemployment was so low it scared the Federal Reserve . . .

Most middle-aged workers were earning 9 percent less than their parents had twenty-five years earlier.

And if more and more women had not gone to work . . .

Most family incomes would not have increased at all during the 1980s and 1990s.[11]

(In the 1950s, both parents worked in one out of five families; today it is one in two . . . Furthermore, the percentage of U.S. men working more than 40 hours per week is 80 percent . . . Japanese men? . . . 62 percent.)

MEANWHILE, THE TOP 20 PERCENT OF SOCIETY . . .
WHICH UNDERSTANDS, WORKS IN, OR INVESTS IN TECHNOLOGY . . .

IS GETTING RICHER . . .
A LOT RICHER.

Growth and wealth . . .

Will be distributed unevenly . . .

As long as a few communities . . .

Pay more attention . . .

To their children's science education . . .

And disproportionately attract the world's best brains.

**While some Americans are getting very rich . . .
many others are not.**

But you do not see a lot of the extreme failures of economic
Darwinism . . .

Because more and more of them end up in jail.[12]

Today the United States, with a population roughly the same as
that of Europe . . .

Jails twice as many of its citizens.

(In 1991, around 936,500 American children had a parent in jail. By 1999
the number had increased 60 percent, to 1,498,800 . . . African Americans
are particularly vulnerable . . . They account for 13 percent of drug
use . . . and 74 percent of those sentenced to prison for drug posses-
sion . . . And it is not getting better; Clinton's administration sent more people to federal prisons than Bush and Reagan combined.)[13]

Educating a population is one of the few things that is consistently more expensive.

As the real price of food, TVs, computers, and vacations drops . . .

The cost of sending someone to college rises.

In 1966, sending a kid to a U.S. private school cost an average of 537 work hours . . .

By 2002 it cost 1,295 hours . . .

(Public school costs also rose from 133 hours to 260 . . .)

It gets ever harder to catch up . . .

And it is particularly hard to catch up if you are not science-literate.

Hispanics are becoming the largest minority in the United States . . .

Followed by African Americans . . .

And there is a lot of emphasis on culture, history, and language . . .

But not much emphasis on science.

IN THE UNITED STATES,
ALMOST ONE-THIRD OF ALL PH.D. SCIENCE AND ENGINEERING STUDENTS ARE ASIAN . . .
TWO OUT OF A HUNDRED ARE AFRICAN AMERICAN OR AFRICAN . . .
ONE OUT OF A HUNDRED IS HISPANIC.[14]

(Meanwhile . . . in California's hospitals . . . the most common name for a baby boy? . . . José . . . By 2050, one out of four Americans could be Hispanic . . . However, during a 2000 gathering of more than 3,200 Latin American scholars in sunny Miami . . . they talked agriculture, Cuba, democracy, economics, justice, law, gender, international relations, migration, race, religion . . . everything except science . . . Sorry, nobody home on that subject.)[15]

As Apollo 13 astronaut Jim Lovell said:
"Houston, we have a problem."

Lack of technical skills leads to lower average wages for African Americans and Hispanics . . .

Less faith in the power of education . . .

And much higher dropout rates . . .

(One-quarter of Hispanics drop out of high school, 13 percent of African Americans, and 7.6 percent of whites.)[16]

This is a problem when close to three-quarters of the children in the Los Angeles County Unified School District are Hispanic . . . and when the United States has become the world's second-largest Spanish-speaking country.[17]

Meanwhile, because tech companies need few people to generate great wealth . . .

And because only the educated tend to get stock options . . .

Salary disparities continue to widen rapidly.

In 1982, a CEO earned, on average, 42 times more than a factory worker . . .

By 1999, the difference was as much as 475.[18]

(Still, being a CEO today is not an easy job . . . Forty of the top 200 U.S. execs were fired in 2000.)

**(By the way, as some regions and ethnic groups end up being left further and further behind or get further and further ahead, they will have ever less of a stake in defending the status quo.
The same people who were surprised when Puerto Rico voted not to become a state in the Union might also be surprised if there were ever fewer than fifty stars in the flag . . .
This is something that has happened to almost every country, except the United States and Brazil.
Often it happens because a few regions generate a disproportionate amount of wealth and feel they pay too much while other regions feel increasingly left out.)**

Technology accelerates trends . . .

Be these positive . . .

Or negative.

A scary world when . . .

The long term . . .

Is no longer measured . . .

In centuries . . .

Or decades . . .

But in years . . .

And sometimes . . .

In months.

It took the telephone **35** years to get into one-quarter of U.S. homes . . .

TV took **26** . . .

Radio **22** . . .

PCs **16** . . .

The Internet **7**.[19]

(And not just businesses are vulnerable . . . It took about nine days for the Berlin Wall to fall and for East Germany to effectively disappear . . . even though East Germany had the strongest economy behind the Iron Curtain . . . and some of the best scientists . . . but no freedom to create and build.)[20]

XI

TECHNOLOGY IS NOT KIND . . .
IT DOES NOT SAY "PLEASE"

In the last decade of the twentieth century, the
global economy bred . . .
Mergers and start-ups on an unprecedented scale . . .
$908,000,000,000 in mergers during 1997 alone . . .
A 47 percent increase over the previous year.
By January 2000 . . .
In one month . . .
There were $244,500,000,000 worth of mergers.

The largest mergers and most successful new companies . . .

Were launched in areas related to . . .

Telecommunications . . .

The Internet . . .

Software . . .

Finance . . .

Mergers driven by the digital revolution.

But there were also some very large mergers and restructurings . . .

Within and between pharmaceuticals . . .

Chemicals . . .

And agribusiness companies . . .

The latter were driven by the ability to read and use . . .

Genetic code.[1]

To get a sense of the massive changes driven by a knowledge economy . . .

Don't just look at computers and telecom . . .

Take a look at the most basic industry on the planet, agribusiness.

(Agribusiness is a concept coined and developed by a wonderful man and mentor, Ray Goldberg, who teaches at Harvard.)[2]

Most of the adults living on the planet today still live off growing, transforming, distributing, or selling food, fiber, and drink.

The ability to genetically alter bacteria, plants, and animals creates extraordinary opportunities.

Instead of growing grain for flour . . .

Farmers are starting to grow . . .

Medicines . . .

Plastics . . .

Fuel . . .

But only those farmers with . . .

Knowledge . . .

And access . . .

To new technologies . . .

Can hope to prosper.

A seed is an instrument designed to execute a genetic program . . .
That transforms soil, water, and sun into wood . . . flower . . . fruit.

Whoever can read and understand a seed's genetic code . . .
Can change it. . . .
So as to reprogram what a plant does . . .
As it grows and reproduces . . .
Which is why an economic endeavor that few paid attention to . . .
Seed production . . .
Became . . .
Very valuable . . .
And very controversial.

The value of a share in Pioneer Hi-Bred International, the premier U.S.
seed company . . .
Increased from . . .

$1.67 IN 1980

TO $4.42 IN 1990

TO $33.75 IN 1999.[3]

When DuPont bought the remaining shares of Pioneer in 1999 . . .
The implicit value of the company was $10 billion.

During the 1990s, a few companies bought up most of the world's independent seed companies.

By 1999, Monsanto and DuPont sold . . .
Over 40 percent of U.S. soy seeds . . .
And almost 80 percent of corn.[4]

And farmers were paying two or three times more for their seeds . . .
(But not for the reason you think.)

In the old economy . . .

The increase in seed prices would have been due to monopoly pricing.

But in a knowledge-driven economy . . .

It was not scarcity driving price . . .

It was knowledge content.

As genetic instructions are rewritten . . .
It gets easier to grow a plant . . .
And cheaper.
And the produce becomes more valuable.

So seeds . . .
Which are a means to accumulate
And transmit . . .
Knowledge . . .
Also become more valuable.

It is not just a matter of growing more with less.

What you grow is becoming very different . . .

Even if you keep calling it corn.

> Dow Chemical has produced corn that becomes a
> biodegradable plastic . . .
>
> DuPont's plants grow a form of polyester that feels like silk . . .
>
> And perhaps thanks to Epicyte, women may eat corn
> to prevent pregnancy.

There is a similar reprogramming going on in other plants . . .

Some bananas and potatoes have been engineered . . .

So that when you eat them . . .

You are vaccinated against various diseases . . . like
cholera and hepatitis.[5]

(New vaccines are important: Over the past fifty years malaria, tuberculosis,
and AIDS have killed six times as many people as those who died in all
wars.)[6]

> The ability to get medicines from plants should not
> surprise us.
>
> After all . . .
>
> Basic sedatives are based on opiates grown in poppies . . .
>
> Many heart-attack remedies come from the digitalis extracted
> from foxglove . . .
>
> Aspirin comes from willow trees . . .
>
> What is different is, we are not just extracting substances . . .
>
> But reprogramming plants to express specific compounds.

(Given that plants naturally produce more than 80,000 different substances,
including vitamins, sugars, starches, and amino acids . . . genetic engineering and farmland could soon substitute for a lot of factories.)[7]

Animals are also changing.

A cute rhesus monkey called ANDi (inserted DNA . . . spelled backward) . . .

Showed that you could insert specific genes into animals very similar to humans . . .[8]

And begin to think about studying Alzheimer's, cancer, blindness, Parkinson's, vaccines, cancer . . .

In a much more deliberate way.

(Start-ups like Ximerex are beginning to engineer pig organs to include human enzymes in an attempt to someday transplant their hearts, kidneys, and livers into our bodies . . . Stem Cells has engineered mice that grow human brain cells to try to cure Alzheimer's.)

Which also raises the specter of eventually engineering human embryos.[9]

This should not surprise us.

There used to be one way of getting pregnant . . .

Now there are more than seventeen . . .

Leading to test-tube babies, surrogate mothers, pregnant grandmothers, orphan embryos.

Which gets lawyers like Lori Andrews to ask . . .

Whether a child conceived from . . .

A donated egg and anonymous sperm . . .

Implanted in a surrogate mother . . .

And brought up by an infertile couple . . .

Has five parents.[10]

But I digress . . . Back to agribusiness.

As these changes occur . . .

Farmers, regions, and countries that do not understand . . .

And help create . . .

Knowledge-driven agribusiness . . .

Are going to get poorer.

One worrisome example is Europe . . .

Where governments and consumers. . . .

Have effectively decided to ban genetically modified seeds . . .

While the European Union continues spending half of its budget . . .

Subsidizing agriculture.

The profits you make in commodities like grain or sugar . . .
Are 3 to 5 percent of sales.

So if the rest of the world does use new seeds . . .

And increases agricultural productivity 4 percent per year . . .

(Not unreasonable given that this was the initial effect of
the green revolution . . .)

The European Union might have to increase its agricultural
subsidies . . .

And spend a further 1 to 2 percent of its total budget . . .

Each year . . .

Compounded . . .

Just to maintain its farmers' current standard of living.

Agriculture is such a big part of the EU budget . . .

That a rapid technological shift could bankrupt the EU as a whole . . .

So either Europe finds a way to face reality . . .

And restructures a lot of farms . . .

Or it blocks imports (and perhaps fights a trade war) . . .

Or it accepts genetically modified plants and animals.[11]

(What Europe decides to do . . . and who it chooses to blame for its woes . . . will have enormous implications for the United States. Already, there are serious disagreements over beef and banana export restrictions. Biotech restrictions will accentuate these tensions and could create significant anti-American sentiment. Never mind Iraq . . .)

The global economy in 2020 will be even more knowledge-driven . . .

Agriculture is simply one example . . .

Of the changes occurring even in the most basic of industries.

But much of the world is way behind in understanding . . .

Adopting . . .

Creating . . .

New technology.

Western Europe is not immune to massive changes . . .

And could easily lose its way if it keeps trying to stop . . .

Key technologies . . .

And keeps bleeding brains to areas like Silicon Valley.

Technology is not kind . . .

It does not wait . . .

It does not say please . . .

It slams into existing systems . . .

And often destroys them . . .

While creating a new system.[12]

(Today's chi-chi economist, Joseph Schumpeter . . . died half a century ago . . . trained as a lawyer . . . coined the term "creative destruction": new products and discoveries relentlessly destroy the old . . . By age 30, he had three clear goals in mind: to become "Europe's greatest lover of beautiful women and Europe's greatest horseman—and perhaps also the world's greatest economist." He claims to have achieved two out of three.)[13]

COUNTRIES CAN EITHER SURF . . . EVER LARGER . . . AND MORE POWERFUL . . . WAVES OF CHANGE . . . OR THEY CAN TRY TO STOP THEM . . . AND GET CRUSHED.

Technology companies are getting very large . . . and very powerful . . .

Microsoft's stock value . . .
Before a judge ruled it a monopoly . . .
Was approaching the value . . .
Of everything Canada produced in one year.

In the first quarter of 2000, AOL Time Warner's market value was greater than everything produced during 1998 by all those who lived in a country like . . .

Albania	Centr. African Rep.	Gambia	Kenya	Moldova
Algeria	Chad	Georgia	Kiribati	Mongolia
Angola	Chile	Ghana	Kuwait	Morocco
Antigua	Colombia	Greece	Kyrgyzstan	Mozambique
Armenia	Comoro I.	Grenada	Laos	Myanmar
Aruba	Congo	Guatemala	Latvia	Namibia
Azerbaijan	Costa Rica	Guinea	Lebanon	Nepal
Bahrain	Croatia	Guinea-Bissau	Lesotho	New Zealand
Bangladesh	Cyprus	Guyana	Liberia	Nicaragua
Barbados	Czech Rep.	Haiti	Libya	Niger
Belarus	Denmark	Honduras	Lithuania	Nigeria
Belize	Djibouti	Hong Kong	Luxembourg	Norway
Benin	Dominican Rep.	Hungary	Macedonia	Oman
Bhutan	Ecuador	Iceland	Madagascar	Pakistan
Bolivia	Egypt	Iran	Malawi	Panama
Botswana	El Salvador	Iraq	Malaysia	Papua N.G.
Brunei	Equatorial Guinea	Ireland	Maldive I.	Paraguay
Bulgaria	Eritrea	Israel	Mali	Peru
Burkina Faso	Estonia	Ivory Coast	Malta	Poland
Burundi	Ethiopia	Jamaica	Marshall I.	Portugal
Cambodia	Fiji	Jordan	Mauritania	Qatar
Cameroon	Finland	Kazakstan	Mauritius	Romania
Cape Verde	Gabon		Micronesia	
				And so on . . .
				And so forth . . .

OR LOOK AT IT A DIFFERENT WAY . . .
OF ALL THE WORLD'S 190 "OFFICIAL" COUNTRIES . . .
LESS THAN ONE-TENTH PRODUCED ENOUGH WEALTH IN 1998 . . .
TO BE ABLE TO BUY AOL TIME WARNER.

Even products that . . .
Are by no means essential . . .
That can be substituted by something cheaper . . .
Or by something made locally . . .
Or by products that are healthier . . .
Can dominate using . . .
Talent, scale, and know-how . . .

AND THE LOGIC OF GLOBAL CAPITALISM IS . . .
THAT ONE MUST DOMINATE . . .
THE WHOLE . . .
WORLD . . .
FAST.

In 1998, Coca-Cola sold 15,800,000,000 cases of various
bottled products . . .
Enough so everyone on the planet could drink a bottle
of its product each week . . .
Many people already drink far more Coke than they do water . . .
But many do not . . .
And this is why the company's 1997 annual report opens
with the statement . . .

THIS YEAR, EVEN AS WE SELL 1 BILLION
SERVINGS OF OUR PRODUCTS DAILY, THE WORLD
STILL CONSUMES 47 BILLION
SERVINGS OF OTHER BEVERAGES EVERY DAY.
WE'RE JUST GETTING STARTED.

(Like a hamburger with that Coke? In 1980, four companies slaughtered 36
percent of U.S. beef; in 1990, it was up to 72 percent; now it is 81 percent . . . You see similar trends in hogs, seeds, grain processing . . . To survive, companies have to scale up globally . . . And in the process, they get
VERY large.)

XII

SLEEPLESS . . . (AND ANGRY) IN SEATTLE

Some think that the only people anguishing about
massive economic change are the colorful protesters
who gather dressed as bioengineered vegetables
outside meetings of the World Trade Organization,
the World Bank, or the World Economic Forum . . .

Wrong . . .

Corporate kahunas are just as concerned and scared
about the rate of rapid change.

The reason so many . . .

Were sleepless . . .

Frustrated . . .

Angry . . .

In Seattle . . .

Was not the World Trade Organization and its policies per se . . .

It was a general sense that the world is moving too fast . . .

And there is no way to catch up . . .

Or have a say . . .

That those who are not market-savvy and technology-literate . . .

Are going to have a hard time even staying in the game.

But just as peoples and regions are scared of the multinational behemoths . . .
Mega-companies are also terrified . . .
That they may not be large enough . . .

Or fast enough to compete . . . **and survive.**

(In the late 1970s, Xerox was *the* tech company, and the FTC was suing it for antitrust . . . It birthed the computer mouse, the point-and-click graphics on your screen, the way to network the Internet . . . But it allowed Apple, Microsoft, Hewlett-Packard, and 3Com to walk away with these businesses . . . In 1998, Xerox stock was trading at $65 per share . . . By the end of 2000, it was around $5 . . . and there were rumors of major restructuring . . . if not bankruptcy.)

Look at the choices faced by pharmaceutical companies, for instance . . .
Creating a new drug . . .
May cost $500,000,000 . . .
And take over a decade . . .

With bets this size, you can lose a third or half the value of your company . . . overnight . . . if you make a mistake.

The genomics revolution could potentially lead to . . .
Anywhere between 4,000 and 10,000 new medicines . . .
Which could require that the pharmaceutical industry spend over . . .
$2,000,000,000,000 on R&D . . .
A sum close to what every person in Latin America produced
in one year . . . (1999)
So it is no wonder that pharmaceutical companies
are merging madly . . .
To try to stay on top of the genetics revolution.

But even if large
companies . . . Grow . . . Merge . . . Dominate . . .

They still find it hard to monopolize, or even attract, the best brains.

(In 1984, the largest U.S. software company was Wordstar
International.)
Some do not need a lot of resources . . . or a lot of time . . . or
people . . . to get very rich.
Small companies can generate a great deal of wealth . . . and get very
large . . . very quickly . . . with ever fewer employees.

HIGH TECH . . . AND A FEW PEOPLE . . . POWER THE ECONOMY[1]

	1986	1996	2006 (PROJECTED)
U.S. JOBS	98,727,000	118,731,000	136,318,000
HIGH-TECH-INTENSIVE JOBS	4.5%	3.8%	4.4%
U.S. OUTPUT CREATED BY HIGH-TECH-INTENSIVE JOBS	5.6%	7.7%	21.1%

A person like Craig Venter can birth entirely new industrial sectors.
Genomics basically did not exist before 1992 . . .
It has generated a few thousand jobs . . .
And an economy about half the size of Chile's . . . in five years.

But caveat emptor . . .[2]

Before you go invest Grandma's pension or your kids'
college fund . . .
Remember that few of the initial leaders of the computer revolution . . .
Prospered, or even survived.
Just like initial IT stocks, genomics companies are
highly volatile and speculative.

A lot of genomic market value disappeared with the
NASDAQ crash . . .

Despite having completed the full gene sequence of
fruit flies, mice, and humans . . .

Established a massive computer center . . .

And accumulated more gene data than anyone else
on the planet . . .

Celera was half as valuable in January 2001 as it had
been in 2000. By 2004, it was sinking.

GENOMICS . . . A BUBBLE? . . . OR A PREVIEW?

COMPANY	STOCK ISSUED	MARKET VALUE (BILLIONS $)		
		JANUARY 20, 2000	MARCH 11, 2000	MARCH 17, 2001
Human Genome	1992	4.4	7.9	5.4
Affymetrix	1996	5.9	6.7	2.1
Geron	1996	0.5	0.9	0.2
Millennium	1996	6.9	10.5	5.1
Celera	1999	5.4	10.5	2.0

In five years, Celera might be Xerox . . . or Wang . . .

Nevertheless, traditional pharmaceutical companies are terrified . . .
By the speed with which a technology . . .
And a new industrial sector can became dominant . . .
Or crash.
Established companies fear that . . .
Genome stocks may simply have overshot their short-term value . . .
And may yet generate monster companies that can take over the old
conglomerates.

(Just ask the folks at Time Warner about that brash little start-up called AOL.)

If new companies . . .
And new economic sectors can appear . . .
In just a few years . . .
Entire economic sectors . . .
And dominant companies . . .
Can also disappear . . .
In a few years.

Few remember An Wang . . .
But through the early 1980s, his machines dominated
corporate America . . .

Because he built the best word processors around.

But the machines never evolved into computers . . .

Nor did they incorporate spreadsheets . . .

And Wang went from nearly absolute dominance . . .

To bankruptcy.

This can happen to the very biggest.

IBM lost its global IT dominance because it did not
realize that . . .

One of its minor contractors had a very powerful idea . . .
(The contractor was called . . . Microsoft.)

The box became a commodity . . .

The value came from the operating system . . .

And in 2004, after very painful restructuring . . .

IBM was worth $140 billion . . .

While Microsoft, even after its close encounter with justice . . .

Was worth $294 billion.

MICROSOFT ITSELF ALMOST FOLLOWED IBM'S PATH . . . IT TOOK BILL GATES A LONG TIME TO FIGURE OUT . . . THAT THE INTERNET WAS NOT A FAD.

(Ironically, Microsoft asked the FCC to block the AOL Time Warner merger if
the company refused to open its instant-messaging protocol.)

That is why Andy Grove, the man who runs Intel . . .

The largest computer-chip manufacturer in the world . . .

Understands the power that technology has . . .

To create and destroy.

So even though he is smart and tough as they come . . .

He chose to title his book . . . *Only the Paranoid Survive.*

(He has destroyed and reinvented his own company several times.)

The great and established . . . are nervous.

Imagine walking into the hallowed offices of "the newspaper of record" . . .

THE NEW YORK TIMES . . .

You can breathe history, meet relatives of the founders, admire old portraits . . .

After all, it was founded in 1851 . . .

A great company, real assets, 13,400 employees, including some of the world's finest journalists . . .

Worth $6.8 billion.

Then wander over to the other coast . . . to Santa Clara, California . . .

And visit a company founded in 1994, with one-seventh the employees of the Times Co. . . .

Some of whom wear snakeskin boots, tight skirts, sneakers, and nose rings . . .

And who are also in the business of providing news and entertainment . . .

They built a little company called Yahoo! . . .

That at one point was worth over $100 billion . . .

Because it had over 120 million users.

(Alas . . . unlike AOL, Yahoo! did not merge and lose 90 percent of its market capitalization.)

This is happening in business after business . . .

The digital economy leverages brains fast.

Sears Roebuck . . . founded 1893 . . . 326,000 employees . . . is worth less than . . .

EBay . . . founded nine years ago . . . 5,100 workers . . . $52 billion market cap . . . *post* NASDAQ crash.[3]

Even if Amazon, eBay, Yahoo!, and other Internet leaders were to
crash . . .
Again . . .
A revolution still took place . . .
The market changed . . .
And an Amazon-like company is likely to dominate retail.

(This phenomenon occurred with the personal-computer revolution.
Bill Gates started programming on an Altair . . .
Commodore 64s were popular for a while . . .
Apple ruled . . .
The fact that the leaders crashed . . .
Did not stop the IT revolution.)

COUNTRIES . . .BUSINESSES . . .

GOVERNMENTS . . .THAT SEEK TO PROTECT . . .TO

MAINTAIN THE STATUS QUO . . .

ARE BOUND TO GET POORER QUICKLY . . . AS

TECHNOLOGY FLOURISHES IN OTHER REGIONS.

(IBM's boss: "Today, even in an industry known for speed, the rate of change
is unprecedented, and it's only being accelerated by the Net. It's fun. It's a
challenge. But if you pause, say goodbye.")[4]

Meanwhile . . .

Back at the genome ranches . . .

Knowledge and databases continue to expand at unprecedented rates.

Even after the

U.S. Patent and Trademark Office

revised the rules on gene patents . . .

Genomics drove around 25,000 new patent requests in 2000 . . .

25 percent more than in 1999 . . .

More than all Internet or computer businesses . . .

Forcing the office to hire 80 new examiners . . .

Who

were

immediately

overwhelmed.

Regardless of the fate of any particular company or country . . .

The revolution is accelerating.

XIII

HIGH TECH . . . HIGH PAY . . .
HIGH MOBILITY . . .

["THE GREAT THING
IN THIS WORLD IS
NOT SO MUCH
WHERE WE STAND,
AS IN WHAT
DIRECTION WE
ARE MOVING."]

OLIVER WENDELL
HOLMES

Seemingly distant and abstract changes in technology . . .

Will change your life . . .

Where you live . . .

What you do . . .

What you earn . . .

Regardless of whether you are a farmer, factory worker, businessman, scholar, or student.

Very few people work in the same company or specialty all their lives anymore.

Your ability to understand and surf waves of change . . .

Will determine how well you do as you change jobs . . .

Every decade . . .

Every five years . . .

Or every few months.

IN SILICON VALLEY . . .
IF YOU ARE NOT STOLEN AWAY BY SOME COMPANY EVERY FEW YEARS (OR MONTHS) . . .
YOU ARE NOT CONSIDERED A HOT PROPERTY.

STABILITY IS A MARK OF SHAME.

(And in this unstable and globalized world, your frame of reference, or even your boss, can change pretty quickly . . . Next time you go to a Kenny Rogers Roasters restaurant, Laura Ashley, or Crabtree & Evelyn . . . remember that you're in a store owned by . . . Malaysian businessmen.)[1]

The only way to make a successful transition from being a corporate man or woman . . .

To being a free agent . . .[2]

Is to learn how to understand and use new concepts and technologies . . .

A skill usually learned in college and graduate schools.

You can see the shift from things to concepts . . .

If you look at how Professor Bob White has restructured the introductory electrical-engineering course at Carnegie Mellon . . .

Students used to focus on resistors, capacitors, inductors . . .

Now they see electricity as a method to accomplish something else . . .

So the first thing they learn is information theory.

Those who do not study suffer . . .

>   In the United States, from 1979 to 1999 . . .

>   College graduates' incomes rose 14 percent . . .

>   High school grads lost 12 percent . . .

>   Those who did not finish high school lost 27 percent.[3]

Even among college graduates . . .

There is a significant difference in incomes between those who understand technology . . .

And those who do not.

There are also significant differences depending on the kind of college degree you get . . .

This difference remains statistically important throughout a lifetime . . .

The technology-literate make, on average, two to three times more than their neighbors.

HIGH TECH . . . HIGH PAY[4]

JOB	U.S. MEDIAN SALARY (2003)
ALL OCCUPATIONS	**$28,392**
Chemists	50,731
Food scientists	56,930
Electrical engineers	71,157
Computer engineers	82,040
Geoscientists	90,750
Engineering managers	99,520

(According to the Bureau of Labor Statistics, each of the five fastest-growing job categories in the United States from 1998 to 2008 is expected to be computer-related.)

How to address educational disparities, digital divides, and technological illiteracy is a hot topic.

All politicians preach an education credo . . .

All promise to spend more money on schools . . .

To make sure better teachers are hired . . .

But few societies have the discipline it takes . . .

To invest and support an effort that will pay off in decades instead of months.

(Over the past fifteen years, the United States' school-completion rate has remained stagnant . . . About one-quarter of all students do not finish high school.)

Various countries have come up with different methods . . .

Let's look at three . . .

South Korea, the United States, and Mexico.

In South Korea...

The country spends 3.7 percent of GNP on education . . .

At least that is what appears in official statistics.[5]

But this wildly underestimates educational expenditures, because . . .

South Korean parents spend almost the same amount on educating their children as the state does . . .

Paying for the best tutors in English, math, and science because . . .

The country is a strict meritocracy, so . . .

At the end of high school, you take a standardized test.

Those with the top scores do not choose where they go to college . . .

They go to Korea University . . .

And they do not choose their major.

Your major depends on your test scores . . .

And even after you graduate . . .

You are still tested.

(A few of South Korea's best students study law . . . But many do not pass the bar . . . to the discredit of themselves and their families.)

It is a brutal system . . .

It is also a system that has catapulted South Korea from feudalism . . .

To technology leader and one of the world's fastest-growing economies.[6]

U.S. parents also spend extra money on their children's education. . . .
Kaplan and Princeton Review make a lot preparing kids for standardized
tests . . .
And then there are also all those extracurricular activities to make the
kids well rounded . . .

    KARATE . . .   SUMMER CAMP . . .   MUSIC . . .
    SOCCER . . .   ART . . . BASEBALL.

(College admissions officers like this . . . or used to. Harvard is now com-
plaining that kids are so overprogrammed and stressed that they look like sur-
vivors of a lifelong boot camp by the time they arrive on campus.)

Yet few high school kids end up able to do globally competitive science
and math.[7]

HIGH SCHOOL MATH . . . WINNERS AND LOSERS

COUNTRY	MATH SCORE
Netherlands	560
Sweden	552
Denmark	547
Switzerland	540
Iceland	534
Norway	528
France	523
New Zealand	522
Australia	522
Canada	519
Austria	518
Slovenia	512
Germany	495
Hungary	483
Italy	476
Russia	471
Lithuania	469
Czech Rep.	466
U.S.	461
Cyprus	446
South Africa	356

(And if you can't pass the test? Well then, why not lower the standards? The
San Francisco Board of Education dropped its graduation requirements for
the class of 2001 after determining that one-third of the senior class would
probably fail . . . Easier to import brains than face political controversy at
home . . . Massachusetts is about to follow San Francisco's lead.)

The number of tech degrees awarded in the U.S. . . .

Has dropped 5 percent per year . . .

Every year . . .

Since 1990.

> (On Boston's elegant Newbury Street . . .
> A plaque on a wonderful old facade . . .
> Commemorates the American Academy of Arts and Sciences (1780) . . .
> A building originally dedicated to research . . .
> And the memory of Alexander Agassiz.
> But now the tenant is Banana Republic.
> As society's priorities changed . . .
> R&D migrated to the suburbs . . .
> Around Route 128 . . . making Weston and Manchester-by-the-
> Sea . . . rich towns.)

As clothing stores flourish in old science temples . . .
U.S. tech education languishes.
A couple of years ago Dorothy and Toto returned to Kansas . . .
To witness a part of a debate as tangled and odd . . .
As any held in *The Wizard of Oz*.
It was part of an ongoing battle . . .
That began in 1925 when a twenty-four-year-old teacher, John Scopes . . .
Was hauled into court . . .
(Actually it was so hot, part of the trial was held outside) . . .
Accused of perverting children's minds with the notion . . .
That monkeys and people shared common ancestors.
It was the first time a trial was broadcast live over the radio to the nation.
Scopes was convicted and fined $100.[8]
He later moved to Venezuela . . .
(A more tolerant place?)
The debate went on..................and on...............
In August 1999 the Kansas State Board of Education . . .
Put Charles Darwin back on a ship and bid him adieu . . .
By deciding that each school district could decide whether or not . . .
To teach the theory of evolution . . .
The origin of the universe through the big bang . . .
The geological age of the planet.[9]

(During 1999, parts of Oklahoma and Alabama were pasting labels on text-books warning that evolution is merely a theory.)

If these extreme examples can occur in the world's most technologically advanced country . . .

Just imagine the challenges faced by entrepreneurs in less sophisticated parts of the planet . . .

And the incentive to move . . . to more tolerant towns.

There are pockets of excellence within the school system . . .
And a lot more creativity than in many countries that emphasize only tests.

But large portions of the U.S. public secondary-school system . . .

Are in real trouble.

Meanwhile academe and business need ever more brains . . .

So a lot of the tech geniuses running around the United States . . .

Were born, and were initially schooled, in other countries.

Maintaining current U.S. leadership in technology . . .

Increasingly depends on . . .

Attracting extraordinarily talented people . . .

From other countries.

(In 1970, about 4.7 percent of the U.S. population was foreign-born; in 2000, it was 10.4 percent.)

Which is one reason U.S. companies are clamoring for increases in H1-B high-tech work visas . . .

More than 600,000 foreign techies will likely get U.S. work permits . . . over the next three years.

(If astronauts leave for Mars on May 6, 2018, the next feasible orbital alignment, it will be because a plasma rocket can get someone to the red planet in 115 days instead of nine months . . . And if the United States gets there first, it will be due to the work of Franklin Ramon Chang-Diaz, a Costa Rican national born to Chinese-Spanish parents, who lived in Venezuela before becoming a U.S. citizen . . . flying the space shuttle . . . and designing a plasma rocket.)[10]

Over 40 percent of all H1-B visas are awarded to Indian nationals . . .

So, in Silicon Valley high-tech, in 1990 . . .

55 percent of Indians had Ph.D.s or master's degrees . . .

40 percent of Chinese . . .

18 percent of whites.

And so by 1998 . . . 2,775 of Silicon Valley's CEOs (one-third of the total) were Indian or Chinese . . . These folks were generating about half as many sales as the billion people in India export over the course of a year.[11]

(Perhaps this is why San Jose is now one of the few U.S. cities with English-speaking cab drivers.)

The United States is not only "borrowing" brains from the rest of the world . . .

It is also running record debt levels.[12]

People focus on the federal budget deficit . . .

But that is not the only place where problems lie.

Companies and individuals are borrowing more and more.[13]

In 1980, private debt was equivalent to 80 percent of what the United States produces in a year . . .

By 2000, it was over 130 percent.

During 1998 and 1999 alone, U.S. corporations borrowed an additional $900 billion . . .

And used $460 billion to buy back their own shares.

(Gee, think this might have artificially inflated stock prices?)

You and I likely also contributed to the problem . . .

The average household now owes more than it earns in a year . . .[14]

And debt keeps piling up because . . .

The amount of stuff the United States buys from the rest of the world greatly exceeds what it sells.

(Imports exceed exports by 35 percent . . . The current account deficit in 2004 will be close to $600 billion.)

Critical budget issues like Social Security and Medicare bills keep getting postponed . . .

To keep growing, the United States has to export more.

(But there is a catch-22—most countries are getting relatively poorer and can buy less because a lot of their capital and brains have migrated to the United States.)[15]

In many developing countries, the educational situation is dire . . .

Many Latin American parents profess a passion for educating their children . . .

But statistics . . .

And government policies . . .

Seem to negate this.

Mexico's income distribution is highly skewed . . .

The richest 10 percent hold 43 percent of the country's total wealth (whereas the richest 10 percent in the United States hold 29 percent of total wealth) . . .

But only 60 percent of the richest children even bother getting a college degree.

> **(And very few of those are science degrees.**
> **Through 1997, those running Latin America's largest media**
> **conglomerate (Televisa) . . .**
> **Boasted that none had a college degree.**
> **As long as most companies remain family-owned . . .**
> **You do not need a good education to succeed.)**

This ethos is transmitted to the country's upper middle class . . .

In many elite schools, the game is "Get away with the least amount of work" . . .

And only 35 percent of those in the upper middle class bother graduating from college.[16]

> Not surprisingly, by the time one gets to the poorest sectors of
> the population, the idea that education can be an instrument of
> social advancement is all but wiped out . . .
> And the government furthers this impression . . .
> By spending a lot of money . . .
> On lousy schools with low standards . . .
> And by avoiding standardized tests . . .
> This makes a meritocracy almost impossible.

Despite everything . . .
Sometimes an exceptional child struggles up the
educational ladder . . .
But the obstacles are often overwhelming.

Rojas is the son of landless peasants . . .
Known affectionately by his diminutive, Rojitas.
Dad never understood Mom's insistence on furthering Rojitas'
education . . .

Even after he was identified as an exceptionally gifted student . . .
Even after earning the best grades in Guanajuato . . .
And receiving a full scholarship to junior high . . .
Dad insisted the boy start working.
But one of Mexico's top college professors . . .
Had employed Rojitas' sister as a maid . . .
Heard the story . . . and insisted he finish high school.

Rojitas applied to college—medical school . . .
With an exceptionally high GPA . . .
But the state university refused to admit him . . .
Because "poor peasants do not become doctors."
They suggested he try law.
Rojitas applied and aced every admissions test . . .
Except one: The socioeconomic test.
The school thought that Rojitas and his family were too poor . . .
So it was unlikely he would stay in school.

When he received his rejection letter . . .
Rojitas was shattered . . .
He was too ashamed to tell the college professor who had
mentored him.
His father screamed . . .
The college admissions board refused to reconsider . . .
So he gave up . . .
Crossed the border illegally . . .
And now lives somewhere in the American South . . .
Working alongside his father . . .
Picking crops . . .
Living on less than the minimum wage . . .
While trying to avoid INS patrols.[17]

What makes Rojitas' case so poignant is that . . .
Mexico's president during this period, Ernesto Zedillo . . .
Also grew up poor . . .
Shining shoes along the U.S. border . . .
And succeeded against all odds . . .
All the way through a Yale doctorate in economics . . .

But when Zedillo returned to Mexico . . .
He became one of the worst education ministers in recent history.
And when he became president . . .
He almost destroyed the continent's oldest university . . .
And favored many things over and above education.
In the first quarter of 2000, Mexico's government spent . . .
$57 million on advertising its accomplishments . . .
$40 million on its science and technology institute . . .
(Which, among other things, funds students abroad . . .)
And $38 million on stores that sell food to the poor.[18]

As the century goes forward, the key driver of economic
prosperity will be technology . . . and scientific literacy.

But many countries just do not get it . . .

Or just cannot keep up.

(Of Mexico's 120,000 schools, only 3,700 were fully wired by 2000.)

Asia gets more and more people trained . . . Latin America, Africa, the
Middle East fewer.

FOREIGN STUDENTS IN U.S. COLLEGES AND UNIVERSITIES[19]
(as a percentage of total foreign enrollment)

REGION	1980–81	1998–99
Africa	12.2	5.3
Europe	8.1	15.0
Latin America	16.0	11.3
Middle East	27.2	6.7
Asia	30.3	56.0

Lack of knowledge leads to poverty . . .
Which means you can send ever-fewer students abroad to study . . .
(Unless you make this a key priority in your ever-smaller budget.)
So it is unsurprising that real economic growth in Africa . . .
Was –0.3 percent per year per person from 1973–96 . . .
And 0.6 percent throughout Latin America.

Only those countries that educate their own or attract the best
from other countries . . .

Are likely to s u c c e e d.

Those who fall hopelessly behind are unlikely to survive . . .

New technologies are generating enormous wealth . . .

They are also creating enormous divisions.

CURRENT ECONOMIC-SHOCK PROGRAMS . . .

ARE JUST A BAND-AID.

It is interesting to read the International Monetary Fund's
take on these trends.

If you cannot get results by increasing real incomes . . .

If national incomes keep diverging . . .

Make up another scale and declare success.

While absolute-dollar income gaps widen . . .

They are peddling a different scale . . .

Called the human-development index.[20]

AS THE FUTURE CATCHES YOU

One way to figure out where a country is going . . .

Is to spend a day at one of the innumerable seminars titled . . .

"The Future of [Country]."

In some cases these seminars focus on IT, genetics, nanotech . . .

But in most cases . . .

Debates and papers focus not on the future but on . . .

History, culture, current political problems . . .

And the smartest get so distracted looking in their rearview mirror . . .

That they end up crashing into the future.

> **Perhaps an extreme example of this . . .**
> **Is Brunei's Technology Museum . . .**
> **Which features "traditional boatmaking, fishing, metalworking,**
> **and goldsmithing techniques—testaments to the ingenuity . . .**
> **(of) earlier generations."**
> **When countries confuse history and technology . . .**
> **They assume the past is a glorious umbrella that will shelter**
> **them from change.**

XIV

THE DIGITAL-GENOMICS
DIASPORA

["WHEN PROGRESS IS
MOVING AS FAST
AS IT IS NOW,
RECALLING ITS
VICTIMS IS
DIFFICULT."]

MICHAEL
LEWIS

A lot of people have little choice . . .

As to where to live . . .

What to work at . . .

How to provide for their kids . . .

For these folk . . .

"Such is life."

But an increasing number of people are realizing that they do have choices . . .

And opportunities . . .

SO THEY ARE
MOVING . . .
TOWARD INDUSTRIES . . .
AND COUNTRIES . . .
THAT TREAT THEM LIKE SHAREHOLDERS . . .
INSTEAD OF SUBJECTS.

Meanwhile, back in Myanmar (formerly Burma) . . .

Anyone owning a PC without government approval . . .

Gets 15 years in prison . . .

Just one of the reasons why . . .

A country with one and a half times as many people as Canada . . .

Did not produce a single U.S. patent in 2003.

Because of the knowledge economy . . .

And the emphasis on human rights . . .

Real, and virtual, migration is growing.

> Worldwide, the number of people living in a country different from
> where they were born . . .
> Has doubled since 1965.
> In Western Europe and North America . . .
> Almost 10 percent of the population is foreign-born . . .
> Many are refugees . . .
> But many are also the best, brightest, and most entrepreneurial . . .
> Who continue concentrating in a few spots.[1]

Many of those wishing to work in the United States . . .

Can now do so freely . . .

Without a visa . . . or visits from the INS.

In a manufacturing economy . . .

You had to travel to the factory and put things together . . .

Or process paper at a specific desk.

In a knowledge economy . . .

You can work at your desk . . .

In your home . . .

In a hotel . . .

In a plane . . .

> Which is why more than a quarter of U.S. workers are
> now part-time, independent, or temps.
> This is particularly true in high-tech . . .
> **Only one out of every three Californians . . .**
> **Holds a traditional 9-to-5 job, working on site.**
> Eighty percent of Sun Microsystems employees work at
> home part-time.
> Average annual salaries for home-only teleworkers are over $50K . . .
> Double the U.S. median.[2]

Those whom you work with . . .

Do not have to know what country or time zone . . .

You are in.

(Bangalore, India, has become the world's second-largest software producer . . . Did you know part of your daily software was made in India? Or that Prodigy and CompUSA are owned by a Mexican?)

MANY U.S. WHITE-COLLAR WORKERS ARE UNDER UNPRECEDENTED PRESSURE . . . BECAUSE FOR THE FIRST TIME . . . THEY ARE COMPETING DIRECTLY WITH WORKERS ABROAD ON A DAILY BASIS.

White-collar workers in other countries . . .

Are leaving their homelands . . .

Physically or virtually (via the Internet) . . .

To work in places where they are treated better.

And . . . often . . . those who leave . . . are the most valuable.

(Over half of all Western European students who earn doctorates from U.S. universities used to stay on for some time after getting their degrees . . . If you do not focus on getting back your best, you soon face a digital diaspora that may devastate economic growth . . . Already half of the total pharmaceutical and biotech research carried out within the U.S. is partly sponsored by foreign corporations, mostly European, because they cannot find the people and knowledge needed within their own countries.)[3]

In many countries, scientists are underpriced, underpaid, and underappreciated . . .

You make more money being an accountant . . .

Not surprisingly, few choose to return.

SCIENCE AND ENGINEERING STUDENTS WHO INTEND TO REMAIN IN THE UNITED STATES POST-PH.D.[4]

COUNTRY	PERCENTAGE
South Korea	33
Mexico	38
Taiwan	44
Italy	47
Canada	48
France	51
Germany	59
U.K.	70
Eastern Europe	72
India	79
China	85

In the measure that citizens go from being subjects of a given state . . .

To shareholders . . .

They can . . .

Abandon . . .

Merge . . .

Divide . . .

Support . . .

Build . . .

Any given flag, border, or anthem.

(Which is de facto exactly what they have been doing.)

THREE-QUARTERS OF THE
SOVEREIGN NATION-STATES THAT POPULATE
THE UNITED NATIONS TODAY . . . DID NOT EXIST
FIFTY YEARS AGO.[5]

(Which is one of the reasons why Olympic opening ceremonies have become so tedious . . . During Australia's 1956 Olympics, 72 teams marched . . . Sydney's 2000 allowed us to nap while 199 teams paraded . . . Just the start of NBC's ratings woes.)

THREE-QUARTERS OF THE WORLD'S SOVEREIGN FLAGS . . . DID NOT EXIST FIFTY YEARS AGO

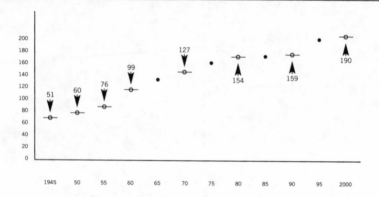

Over the past decade . . .

The world has bred an average of . . .

Three new states per year . . .

Which implies that the next generation . . .

May have to memorize the capitals and flags . . .

Of more than 100 additional countries.[6]

(Many countries are highly fragmented within in terms of language, ethnicity, and culture . . . Just for starters, Brazil, Indonesia, and India host 210, 670, and 850 distinct languages and dialects, respectively.)

So far . . .

Most political scientists . . .

Have assumed that the reason so many countries are D I V I D I N G . . .

Is because of religious, ethnic, cultural, or linguistic cleavages.

But these are divisions that have usually existed for centuries . . .

So their explanation fails to address two basic questions . . .

Why the sudden impulse to secede?

And why are so many now successful?

Part of the answer is that . . .

Technology empowers individuals . . .

And makes it hard to prevent democracy . . .

Which in turn allows people to ask . . .

Do I really wish
to remain a shareholder
of this state

(Increasingly often the answer is

NO!

but thanks anyway.)

If what really counts is having smart people . . .

(And remember . . . working with these folks can be pretty fruity. The guy who built the Internet got his doctorate studying the auditory cortex of cats.[7] Kary Mullis got the Nobel prize for inventing a genetic engineering technique called PCR . . . Then he wrote *Dancing Naked in the Mind Field*.)

And the most valuable product is the ideas they generate.

The way you accumulate wealth is . . .

To get permission to continue interacting with these people.

They have to be willing to share their time and space . . .

But . . .

> When you . . .
>
> Open borders . . .
>
> Push free trade . . .
>
> Build a knowledge economy . . .
>
> Preach universal human rights . . .
>
> Allow free and open communication . . .

You had better also remember . . .

That those who now have a choice . . .

May leave . . .

And that means . . .

GOVERNMENTS HAVE FAR LESS LEEWAY . . .

IF THEY ABUSE THEIR POWER . . .

IF THEY IGNORE CITIZEN-SHAREHOLDERS . . .

THEY WILL TEAR APART THE STATE.

After decades of dictatorship, Spain finally got rid of
Generalissimo Franco . . .

Opened its borders, established democracy, joined
the European Union.

It established a joint program between its Royal College
and Harvard . . .

Got King Juan Carlos to fly to Boston to open a new building . . .

And sent some of its best and brightest to Cambridge,
Massachusetts, to get master's degrees and doctorates.

Great stuff . . . but not enough . . .

Not one of these graduates has gone home (yet).

One of Spain's distinguished scientists explained . . .

"We have too many people . . . There is no space for them."

(Wonder if this attitude may have something to do with the fact that
it takes nine times as many Spaniards as Frenchmen to patent something
in the United States.)

Meanwhile, Spain keeps falling apart . . .

Basques, Galicians, Catalans . . .

See ever less reason to support the whole.

After all, under EU rules . . .

A new border does not change their passport, currency, security . . .

They can work anywhere, travel at will, run their own affairs without inter-
ference from Madrid.

And for the first time in centuries . . .

They have a real choice.

(The smartest and most successful countries never forget this lesson.
Singapore's founder, Lee Kuan Yew, ends his memoirs with, "Will Singapore
the independent city-state disappear? The island will not, but the sovereign
nation . . . could vanish.")

India is another example of a State that might fall apart.
The world's largest democracy . . .
It built six extraordinary technical schools . . .
That train some of the world's best engineers and scientists . . .
But many graduates find it very hard to start a new company . . .
Because of a myriad of traditions and regulations.

(For instance, the 1888 Telegraph Act forbids using telecom for a profit . . . Gee, do you think that might put a crimp in e-commerce?)

So the brightest and most ambitious leave . . .
And power Silicon Valley with "ICs" . . .
(Not integrated circuits . . . Indians and Chinese.)
ICs are senior executives in over one-quarter of the area's high-tech firms . . .
With about $17 billion in sales in 1998.[8]

(Equivalent to half of all of India's exports.)

(Meanwhile, the billion-plus people left in India . . .
Obtained a single patent in the United Kingdom in 1999 . . .
And exported about $32 per person.
A few regions grew and became competitive . . .
Most did not . . . India, if it wishes to stop breaking up . . .
May want to pay attention to these trends.)

A country's job, a government's task . . .

Is to grow, develop, keep, attract talent . . .

And make sure this talent creates and protects
new knowledge . . .

That can launch new companies . . .

Otherwise, there is no economic growth . . .

No funding for the arts, no welfare, no stability.

(A government's task gets ever harder as key people are freer to choose where they pledge allegiance, hold their assets, build a real or virtual community. The technologically literate have a global passport and their citizenship is now a market.)

Education . . .

Democracy . . .

Technology . . .

Competitiveness . . .

Individual economic opportunity . . .

All these overused . . .

Seemingly *trite* words . . .

Have become matters of national security . . .

(During 1999, there were twenty-seven major wars on the planet . . . Twenty-five were civil wars . . . mostly in poor countries. Large-scale ethnic diversity does not increase the likelihood of civil war. What does lead to destroyed states is: a lot of commodities, substantial brain drain, a lot of uneducated and unemployed teenagers, and a declining economy.)[9]

Nations and civilizations . . .

Do not prosper, or even survive, very long . . .

If they can't provide the fundamental pillars . . .

Of a knowledge-based economy.

(But don't you worry—they leave wonderful archaeological ruins to tour in a rickshaw or on camels . . . Just go on and keep ignoring scientific revolutions.)

XV

TIME WARP

[
"GETTING A
GENOME SEQUENCE
HAS NEVER BEEN
AN END . . .
JUST A START."
]

CRAIG
VENTER

If, heaven forbid, . . .

Your young children died in an accident . . .

You would probably do anything in your power to try to bring them back . . .

Some parents are exploring this option.[1]

Almost any species can be cloned today . . .

(Although the success rate is quite low . . . below that of in-vitro fertilization attempts.)

In the United States, it is illegal to use federal funds to clone humans . . .

But it is not illegal to clone a human . . .

(Except in California, Louisiana, Michigan, and Rhode Island.)

Nor is it illegal in Singapore, Russia, Brazil, China.

And if you combine desperate customers . . .

Rapidly evolving and highly decentralized technology . . .

And the moniker of "the first scientist to clone a human" . . .

The incentives are too great to stop this from happening . . .

Very soon.

(Pity Marilyn is no longer alive . . . Remember those multicolor single-image silkscreens Andy Warhol used to make? . . . They might have become a reality.)

A clone is not a photocopy . . .

You would be a different parent . . .

Grandma might not be around to spoil the kids . . .

Teachers, friends, and foods would differ . . .

Cloned kids would be strikingly similar to and strikingly different from their predecessors.

> But you would get a part of your child back . . .
>
> (Think of a genetically identical twin, but several years or decades younger.)
>
> Clones are going to teach us a lot about what is nature and what is nurture . . .
>
> About how much our genes govern our thoughts and actions . . .
>
> Likely, we will be surprised by just how different . . .
>
> Our clones will be.

Cloned kids are just the first step in a very different world . . .

> If you can rebirth your kids' DNA . . .
>
> The same can be done for your parents . . .
>
> Your grandparents . . .
>
> Yourself.
>
> (Or you can merely rebirth your cells or your organs.)

The so-far absolute notion . . .

Of a linear and finite . . .

Decades-long life . . .

Is about to enter a time warp.

While this may surprise you and me . . .

It is old news to scientists like James Watson.

In **1971,** just as the first in-vitro fertilization techniques were being developed . . .

Watson argued: "Given the widespread development of the safe clinical procedures for handling human eggs, [human] cloning experiments would not be prohibitively expensive. They need not be restricted to the super powers. All smaller countries now possess the resources required for eventual success."[2]

He was right. It took about three billion dollars and over a decade for the public genome project to ramp up and sequence a human.

Today it costs less than $30 million to sequence a mammalian genome . . .

And the cost is falling every week.[3]

There are now at least ten labs that have the capacity to sequence and decode a full human genome in a few months.

SCIENTISTS AND COMPANIES ARE RACING FLAT-OUT . . . TO DECODE . . . UNDERSTAND . . . REWRITE . . . PORTIONS OF VARIOUS GENOMES.[4]

**The Christian moral and ethical system is ill-equipped
to address some of the choices and dilemmas created by the
genomics revolution.**[5]

(I say this as a quasi-Catholic in a Protestant family.)

We may all want to pay some attention to the beliefs and
consequences . . .

Explored by religions like Hinduism and Buddhism . . .

Where reincarnation remains a central tenet . . .

(Among other things . . . where one ends up . . . depends on how one acts in
one's current life.)

Karma and reputation matter when your grandkids or great-
grandkids . . .

Will be the ones deciding whether to bring part of you back . . . or not.

If the idea of cloned kids, grandparents, selves seems
too far out . . .

Reflect a little on what already seems normal.

A wonderful Princeton professor wrote a book titled *Remaking
Eden,* which argues that . . .

The world changed July 25, 1978 . . .

When Louise Joy Brown was born . . .

The first child ever conceived outside a mother's womb . . .

**Created not by a physician . . .
But by an expert in mouse embryos . . .
Someone trained in animal genetics . . .
(A preview of coming attractions?)**

There were immediate and widespread demands to stop
"test tube babies" . . .

Tabloids had a field day . . .

But to the relief of many anguished couples . . .

Hundreds of thousands of babies have now been born in vitro . . .

And today three-quarters of Americans think this procedure is OK.

(Meanwhile, Louise, like any normal teenager, turned 18 flipping burgers and studying to be a nurse.)

Technology kept pushing . . .

On March 28, 1984, a once-frozen embryo . . .

Was born Zoe Leyland . . . a pert Australian.

Today, cryopreservation is common . . .[6]

It means someone can be born . . .

Decades or centuries after parents die.

(A lot of people tried to adopt the two embryos conceived by Mario and Elsa Rios who were orphaned after a plane crashed in 1983 . . . Perhaps this was simply an example of Good Samaritanism . . . or perhaps it might have something to do with the $8 million estate they left behind and the lack of a living will . . . The embryos are in legal limbo, floating in liquid nitrogen in Melbourne's Queen Victoria Medical Center.)

The world changed yet again February 23, 1997 . . .

When a lamb named after Dolly Parton was born in Scotland.[7]

In this case there was no conception, in or out of a female womb . . .

No freezing of an embryo.

Dolly was cloned from an adult cell . . .

This implies we could someday create copies of ourselves from any cell in our body.[8]

(WHICH MAKES

GENESIS 1:26

A LOT MORE POIGNANT:

"LET US MAKE MAN IN OUR IMAGE,
AFTER OUR LIKENESS.")

We are now beginning to alter pig genes . . .

To make them more human.

There are good reasons to do so . . .

There are a lot fewer organ donors than there are really sick people.

In the United States over 70,000 people are on transplant wait lists . . .

So heart valves are often replaced with pig tissue . . .

But using a different person's body parts, never mind a different species . . .

Greatly increases the possibility of rejection . . .

And requires massive doses of immunosuppressants.

However, if you remove some surface proteins from pig tissue . . .

Humanize them . . .

The chances of rejection decrease.[9]

These techniques lead to interesting questions . . .

As we substitute human genes for pig genes in an animal . . .

How far are we humanizing it?[10]

(Of course George Orwell realized how similar pigs can be to some humans a long time ago . . . Just reread *Animal Farm*.)

As of February 12, 2001 . . .

We know what the human genome consists of . . . (with an accuracy of
99.96 percent).[11]

(Actually part of it was leaked a day early because a British paper, *The
Observer,* broke the news embargo . . . proving yet again that journalism and
ethics do not always splice.)

26,588 genes . . .

0.9	percent run our immune system . . .
2.9	prevent tumors . . .
3.3	allow cells to communicate with one another . . .
5.0	build cells . . .
10.2	make enzymes for chemical reactions . . .
13.5	run the cell nucleus . . .

And . . .
We do not know what the other 41.7 percent do.[12]
(Another 12,000 bits could eventually be classified as genes . . . But the
end count is likely around 30,000.)
You can put them all on a gene chip . . .
The size of a penny.

Among the lessons from the genome . . .
Men are in trouble.
Women have two X-chromosomes . . .
Men have one X and one Y.
Ys are a third as large as Xs . . .
Recessive . . .
Repetitive . . .
Twice as prone to mutations.

In the words of a Johns Hopkins professor . . .

"The Y chromosome
is really pitiful."

(Maybe this is why men . . . fight more . . . live less . . . think they run things?)

But pity mankind as well as men . . .
Only 300 of our genes . . .
Out of a total of 26,588 . . .
Have no counterpart in mice . . .
And about 85 percent of the two species' gene sequences are identical.

WE ARE STARTING TO UNDERSTAND JUST HOW INTERWOVEN ALL LIFE ON THE
PLANET REALLY IS . . . AND SOME ARE
BEGINNING TO LOOK FOR "MODEL GENOMES" ACROSS
SEPARATE SPECIES . . . EVEN
"LOWER" SPECIES.

One of the great gene scientists, Michael Eisen, recalls
that when he was in sixth grade . . .
He and his fellow monsters drove a substitute teacher
over the edge . . .
To the point where she started screaming . . .
"You are all worms . . . just a bunch of worms."
Little did she know how right she was . . .
Someday research on the genes of a rat, fly, or worm . . .
May save your life.
Because basic cells and genes are conserved across species . . .
Unless there is a good reason to change.

Friedrich Nietzsche . . .
And many wives understand . . .

"We have
made our way
from worm
to man, but
much of us is
still worm."

(Thus Spoke Zarathustra.)

Perhaps now you have an inkling of the unease felt . . .

By those folks who had just figured out that . . .

The universe does not revolve around the earth . . .

The planet is not flat . . .

Time is relative . . .

And it warps.

But fear not . . .

What makes us special is not the number of genes . . .

Or the fact that we share many of these with worms, plants, bacteria.

What is particular to humans is the complexity with which we network . . .

Our biological selves.

An average gene encodes three different proteins through alternate splicing . . .

It is how our genome interacts that ends up making a huge difference.

(Just as computers really did not come into their own until they were joined through the Internet.)

Stanford's Stuart Kim explains this concept using classic books.
Worm genomes are like Dr. Seuss:
"Yertle the turtle was king of the pond. A nice little pond. It was neat. The water was warm. There was plenty to eat."

The same alphabet and language can create great complexity using short sentences.

Human genomes are closer to Dickens . . .
Where words modify and build by creating tension with each other . . .
"It was the best of times, it was the worst of times, it was the age of wisdom, it was the age of foolishness, it was the epoch of belief, it was the epoch of incredulity."

Now that we have the first great map of what it means
to be human.

What we do not know . . .

Is where this information will lead us . . .

How we will choose to modify human DNA . . .

Whether these alterations will take place only
within our body . . .

Or also in those of our children and all subsequent generations . . .

Whether we will be rational enough to keep genes and
war separate.

Australians attempting to make mice infertile . . .
Inadvertently created a deadly mouse virus . . .
Demonstrating yet again just how important it is that . . .
We fight for and reinforce the 1972 Biological Weapons Convention.[13]

We have to remember the grievous mistakes we made . . .

When we tried to judge each other . . .

To stand above another . . .

To fight one another . . .

According to race or genes.

(You do not really dream of being engineered a blueblood, do you? . . . Think
Prince Charles.)

Historian William Manchester described the transition from the Middle Ages into the Renaissance . . .

As a transition within *A World Lit Only by Fire*.[14]

In a sense, today's medicine, agriculture, manufacturing, and data processing . . .

Are still very primitive . . .

Full of superstitions, assumptions, and false science.

We still live in a world mostly ignorant of the basic genetic alphabet . . .

With machines much dumber than we are . . .

And remain convinced we are the only rational and intelligent beings in the universe.

BUT WE ARE BEGINNING TO GLANCE AT A NEW WORLD . . . AND ARE BEGINNING TO REALIZE HOW PROFOUNDLY DIFFERENT OUR CHILDREN'S LIVES WILL BE.

The rule in bio labs used to be:
One grad student = One gene = One Ph.D.
It took Celera nine months to map the human genome . . .
Then six months to do the mouse.
Within a year all our genes will be on chips . . .
Meanwhile, my fourth-grader, Diana, and my second-grader, Nico . . .
Have taken the DNA out of an onion . . .
And can help run it through a sequencing machine . . .
To identify genes.

As gene maps open a New World . . .

Genomics is ground zero.

Sam Broder . . .

One of the world's great brains . . .

AIDS researcher . . . ex-head of the National Cancer Institute . . .

Who could have worked in any number of NGOs, hospitals, corporations, or universities . . .

Chose to become Genome Company's chief medical officer . . .[15]

Because that is where he thought he had the best chance . . .

Of identifying proteins in blood, in spinal fluid . . .

To diagnose cancer before symptoms appear . . .

While helping develop cancer-specific vaccines.

Broder worked with a soft-spoken gentle giant . . .

Ham Smith . . .

Who won the Nobel prize in 1978 . . .

(And whose dream, until a couple of years ago, was to be able to buy a small farm.) Now he has the farm and wants to build artificial chromosomes.

At science conferences, you might run into a punkish-looking computer genius, Gene Myers . . .

And Mark Adams, a guy who looks like he is still in high school . . .

But who is able to assemble gene sequences like they are Legos . . .

And who may someday outshine Watson and Venter.

(And of course there is Claire Fraser . . . who quietly gardens and tends to three giant poodles who think they are human . . . during the few hours that she is not running the world's premier microbial gene sequencing institute, TIGR.)

Venter's small group of privateers . . .

Took on the governments, companies, and universities of the world . . .

With a project that seemed impossible . . .

And accelerated everyone's timetable by years.

It is because of their drive and ambition that we now have two rough maps of our species . . .

These maps will be every bit as important as those drawn by Columbus, Cook, and Galileo . . .

Now we have to refine the charts and begin navigating.

AS THIS BOOK ENDS . . .

YOU SHOULD REMEMBER THE CONCLUSIONS REACHED BY THE AUTHOR . . . OF A MODESTLY TITLED BOOK CALLED *HISTORY OF THE WORLD* . . .[16] AFTER ALMOST ONE THOUSAND PAGES . . .

HE CONCLUDES TWO THINGS . . .

> HISTORY CHANGES FASTER
> THAN ONE MIGHT EXPECT . . .
> AND HISTORY CHANGES SLOWER
> THAN ONE MIGHT EXPECT.

In the 1960s, many expected flying cars, floating cities, large space stations . . . by 2001.

They did not see the impact of pervasive instant, global, almost free networking . . .

Of a massive proliferation of mostly nonviable states.

In 1900 . . . or before the Berlin Wall fell in 1989 . . .

It was hardly obvious that the United States would become the world's sole hegemon.

(One of the best-selling books in the 1980s? . . . Ezra Vogel's *Japan as Number One: Lessons for America.*)

Nor is it obvious where Japan, China, Russia, Brazil, Singapore, the United Kingdom, or the United States will be at the end of the twenty-first century . . .

(The current U.S. mantra: "U.S. as Number One: Lessons for the World" . . . may sound a little old in 100 years.)

What our world looks like in fifty or a hundred years . . .

Depends to a significant extent . . .

On our ability to adopt and adapt . . .

The ethical, political, and economic challenges . . .

Of the digital-genomics era.

As we learn more about life . . .

Mendel, Darwin, Watson, Crick, Venter. . . .

Will be figures every bit as important as Edison, Einstein, Ford, the Wright brothers.

(Each built, or inspired, the industries that make the United States the dominant economy and superpower today.)

What they have taught us and produced is changing each of our lives . . .

How we work, live, and think.

You can stand on the sidelines and assume fate will guide things . . .

(God willing . . . Si Dios Quiere . . . Insha'Allah . . . Shikatta ga nai . . .)

Or you can help yourself, your family, your company and country navigate . . .

This wondrous and scary adventure.

I hope I have convinced you of the need to do the latter.

(And you are now better prepared than most politicians are to face this task . . . When the Pennsylvania legislature was debating a bill to ban cloning, a mischievous gene ethicist, Art Caplan, asked various lawmakers where your genome lies . . . one-third said the brain, one-third said in the gonads, one-third sort of got it right.)

We are about to go over a waterfall . . .

And in a few centuries . . .

February 12, 2001, will represent the divide . . .

Between an era before humans had mapped genes . . .

And the post-genomic era.

THE END . . .

(ACTUALLY, JUST THE BEGINNING.)

POSTSCRIPT

I apologize for simplifying so many debates and concepts. My objective is not to teach you everything you need to know about technology but, rather, to start a debate. I hope this will feel like a Chinese meal and leave you hungry to read more, which is why I included detailed endnotes.

It took four years of research to write this book. I began, and have almost completed, the manuscripts of three "serious" books as well as a variety of cases, articles, and chapters before writing this one. But I finally decided that I had to finish this book first, as an introduction to a more in-depth debate about science and the future of the nation-state. My agent, Barbara Rifkind, provided advice and support. She made this a reality by bringing the project to Random House, where an extraordinary editor, John Mahaney, and Shana Drehs, guided the book.

Many people took a lot of time to educate me about various technologies and about how the world works. Harvard was an extraordinary venue. Those teaching and researching in the Center for International Affairs helped me understand what keeps countries together, and what tears them apart. Three people, Derek Bok, Jorge Dominguez, and Bob Putnam, started me on this adventure by allowing me to join an extraordinary group of CFIA Fellows (I am particularly indebted to Russ Howard, Derek Offer, Pete Bunce, Marta Lucía Ramírez, Imelda Cisneros, Diego Hidalgo, and Luis Fernando Ramírez). In South Korea, many people went out of their way to be helpful, particularly Professors Byung-Kook Kim and Young-Jin Kim, as well as Minister Dal Ho Chung. Stanley Kao, a patriot, taught me about surviving and thriving in Taiwan. Mahathir Mohamad and Noordim Sopiee taught me a lot about Malaysia and why things are not as easy as they look. Gloria Arroyo and Bobby Romulo did the same in the Philippines. Australia's Phil Scanlan and Alan Carroll taught me to understand Asia better.

At the David Rockefeller Center for Latin American Studies, John Coatsworth, Steve Reifenberg, and June Erlick helped create a home instead of an office. Tim Stumph killed whole forests printing out version after version of book drafts. Special thanks to Professors John Womack,

Marcelo Suarez Orozco, Wayne Cornelius, Abe Lowenthal, and Mala Htun. Professor Otto Solbrig, one of the world's great scientists, provided encouragement, knowledge, and a home in Harvard's biology labs. Of course, without Neil and Angelica Rudenstein, the center would not exist . . .

Five Harvard Business School professors were especially influential on my thinking and spent hours going over charts and ideas. Ray Goldberg, the father of agribusiness, taught me about hope, hard work, and constructive change . . . Bruce Scott about competitiveness, countries, and having the patience and talent to teach counterintuitive ideas . . . George Lodge about sailing, ideology, development, and kindness. Debra Spar always had a smile and encouragement as well as a sharp sense of how to improve arguments. And Jonathan West helped me conceive and launch the life-science project.

Joan Bok and Al Houston asked me to join the genetics advisory council of Harvard Medical School, which provided an extraordinary opportunity to learn from professors like Phil Leder, Connie Cepko, George Church, and Hidde Ploegh. Robin Blatt was kind to ask me to become a contributing editor of *The Journal of Biolaw and Business*.

Throughout Mexico and Latin America, many people taught me how to do things right and how to screw up a country. I thank both groups for valuable lessons. But I'll mention only those who taught me that Mexico is and will remain a great country. Foremost, my father, Antonio Enriquez Savignac. Without him, there would be no Cancún, Ixtapa, Los Cabos, Loreto, or Huatulco. He exemplifies the ideals of this book—that one man, willing to learn and create, can make a huge difference to a country and improve the lives of millions. And that what counts is not what one says but one's ability to execute. He was twice elected Secretary General of the World Tourism Organization, and spent a decade building Cancún's successors throughout the world.

Manuel Camacho and Marcelo Ebrard taught me courage—what it means to fight to keep a country together against all odds. We were able to stop a war in Chiapas. Many other Mexicans from various parties, schools, companies, and publications also taught me about what it takes to get ahead, to remain competitive against all odds: Adolfo Aguilar, Oscar Arguelles, Jorge Castañeda, Pedro Cerisola, Oscar Elizundia, Ramón Alberto Garza, Nacho Marván, Enrique Rangel, Rubén Ojeda, Jesús Silva Herzog, Luis Sanchez, José Sarukán, Julio Scherer, Leopoldo Solís, and Rene Solís.

A significant portion of the research and ideas contained in this book came from Rodrigo Martínez. Alison Sander always had time to teach me what she was finding as a successful entrepreneur and as the global fore-caster at Boston Consulting Group, how to synthesize complex trends, and how to look at the world in a different way. Tom, Cynthia, Jay, John, and Ann Schneider provided ideas, edits, and support. Alan Stoga and David Quilter patiently questioned. Jonathan Slonim, Mary Schneider, Gaye Bok, and Antonio Enriquez Cabot revised the manuscript with a sharp pencil and a lot of laughs and a few tequilas. Paul Davis and Seed Partners make some of these ideas reality.

Some extraordinary men and women took the time to convince me about the primary importance of education and technology: Claire Fraser, Craig Venter, Jose Maria Figueres, Rod and Nancy Nichols, Timothy Ong, Al Gordon, Andrew Cabot, and especially David Rockefeller. His work with the Americas Society has pushed the private sector to get far more involved in education, an effort led by Peggy Dulaney, Roberto Paulo Cezar de Andrade, Fernando Romero, Patricia Cisneros, and Luisa Pulido. Matt Neville, Sam Bodman, and Ken Burnes shared their expertise on how to integrate research and business

For generations, my family have been wanderers, to and from America, Asia, and Europe. That the younger generations choose to strive is due to extraordinary leadership: John and Elizabeth Moors Cabot made the world better in embassy after embassy; Louie and Muffie Cabot led in business, politics, and thought; Jane Bradley is an example of what women can do; Lewis Cabot and Lisa Lyman reinforced the need to always look for beauty. Above all, John and Carroll Cabot quietly provided shelter, guidance, and example, in the process ensuring all the rest of us can take risks and strive. Thank you.

Finally, Diana, Nicolas, and Mary didn't always understand why I would get up early, put on sweatpants, and head for the attic to build moats of paper and write day and night. It was hard to explain why I traveled so much, and why our house always seemed full of scientists, journalists, politicians, businessmen, and historians. They never lost their smile and humor when I called from Argentina, Australia, Brazil, Brunei, Cambodia, Chile, Egypt, Japan, South Korea, Mexico, Nepal, Peru, the Philippines, Singapore, Taiwan, Tanzania, Thailand, various parts of Europe, the United States, and Canada. Each of these countries and their extraordinary people taught me time and again how important it is to be a true epicurean . . . a citizen of the world.

Cambridge, Massachusetts

March 2001

N O T E S

Chapter I: Mixing Apples, Oranges, and Floppy Disks . . .

1. "Stable Germline Transformation of the Malaria Mosquito *Anopheles stephensi*," *Nature* 405 (2000): 959–62.

2. Britain was the first country to allow this. The measure passed in the House of Lords 212 to 92, despite the opposition of various religious leaders. The clones can grow only to 14 days. The objective is to create human stem cells (undifferentiated cells that can grow into any tissue), in an attempt to treat leukemia, Parkinson's, and cancer.

3. Of course not everyone gets excited by new maps. During the presentation of the human genome at the American Association for the Advancement of Science meetings . . . folks were scrambling to get their hands on the first printed genome maps . . . One enterprising reporter snagged her copy . . . ripped it open . . . spread it out on the floor . . . and used it as a quilt to lie on during the press conference.

Chapter II: The 427:1 Gap

1. David Walton, "Infectious Diseases: Looking at Medical and Social Justice," *DRCLAS News,* fall 2000.

2. People in every occupation go bankrupt. See Elizabeth Warren, Teresa A. Sullivan, and Jay Lawrence Westbrook, *The Fragile Middle Class* (New Haven: Yale University Press, 2000). Serious medical problems are behind 40 percent of bankruptcies and divorce behind 20 percent.

3. Rick Wyatt researched the various flags of the Confederacy. You can look at the whys and hows at http://fotw.vexillum.com/flags/us-csa1.html. Mark Sensen and others did an analysis of the dozens of Soviet flags: http://fotw.vexillum.com/flags/su.html.

4. The world's premier economic historian-statistician is Angus Maddison. *Monitoring the World Economy, 1820–1992* (OECD, 1995).

5. Take a look at the Model T club . . . www.modelt.org/thecars.html.

6. David S. Landes has worked extensively on the differences between rich and poor; see, for example, *The Wealth and Poverty of Nations: Why Some Are So Rich and Some So Poor* (New York: Norton, 1998). Jeffrey Sachs has also worked on the differences between the India and China of yesterday and of today (see www.cid.harvard.edu).

7. Craig Canine wrote a good book on the transformation, mechanization, and consolidation of U.S. agriculture: *Dream Reaper: The Story of an Old-Fashioned Inventor in the High-Tech, High-Stakes World of Modern Agriculture* (New York: Knopf, 1995).

8. There is a healthy debate as to how representative these tests are and what they mean. Numerous editorials have been written in publications like *Science*. For an official position, look at U.S. Department of Education, National Center for Education Statistics (NCES), *Pursuing Excellence: A Study of U.S. Twelfth-Grade Mathematics and Science Achievement in International Context,* NCES 98-049 (Washington, D.C.: U.S. Government Printing Office, 1998).

9. The Big Mac Index reflects the cost of purchasing this hamburger at McDonald's restaurants throughout the world. It was created in 1986 by *The Economist* to look at purchasing-power parity. These figures are calculated by *Reforma*: www.reforma.com/flashes/negocios/big_mac/. Some of the actual differences between countries may be even larger than they appear in these numbers because the average manufacturing wage is divided by the cost of a Big Mac, and a burger costs $2.09 in the United States and $1.10 in India.

Chapter III: The New Rich . . . and the New Poor

1. The U.S. Bureau of Labor Statistics is an endless fountain of data to drown in. This table comes from *International Comparisons of Hourly Compensation Costs for Production Workers in Manufacturing*: www.bls.gov/news.release/ichcc.t02.htm. Want more detail? Check out ftp://ftp.bls.gov/pub/special.requests/ForeignLabor/ichcctab02.txt.

2. Federal Reserve Bank of Dallas, 1997 Annual Report. There are some amusing comparisons between 1900 and 1998 data in *Time*'s special issue on the most influential people of the twentieth century. www.time.com/time/time100/index.html.

3. *World Development Report 1982*, pp. 114–115 and *World Development Indicators 2000*, pp. 188–89, World Bank. For 1960, the percentages of the distribution of the GDP are the average of the countries classified as: low-income and middle-income economies; industrial market and nonmarket industrial economies. The percentages don't add up to 100, but they appear this way in the source.

4. Seth Godin has talked and written about these trends at Cap Gemini Ernst & Young seminars. See also his book *Permission Marketing* (New York: Simon & Schuster, 1999).

5. Kenichi Ohmae has done a series of studies of this phenomenon.

6. One of the great minds on competitive analysis is Harvard Business School professor Michael Porter. He has looked at flowers as well as a myriad of other industries in various books, including *The Competitive Advantage of Nations*, *Competitive Advantage*, and *Competitive Strategy*.

7. A lot of people have written on this phenomenon of a new economy; one of my favorites is *Wired* editor Kevin Kelly. Many of the ideas in this chapter come from his book *New Rules for the New Economy* (New York: Viking, 1998). Peter F. Drucker has also been detailing these changes for decades, starting with *The End of the Economic Man* (1939). See also his *Post-Capitalist Society* (1993).

8. Robert Metcalf, founder of 3Com, argues the value of a network is proportional to the square of the number of people in it.

9. Attempting to regulate these decentralized networks is certain to give bureaucrats ulcers—or worse. See, for example, Peter Maas, "Silicorn Valley," *Wired*, September 1997: 131–38.

10. You can read about SETI in Howard Rheingold's "You Got the Power," *Wired*, August 2000: 176. Or go directly to www.seti.org/.

11. See www.fightaidsathome.com.

12. Federal Reserve Bank of Dallas. 1997 Annual Report: "Time Well Spent: The Declining Real Cost of Living in America."

13. While the Internet has certainly changed the nature of shopping, it is still hard to go up against merchandising behemoths like Wal-Mart. The skills and techniques that allowed

Amazon to dominate in one arena now have to be honed in others to achieve that incidental part of business . . . making a profit.

14. Value in January 2000. By May, the value had almost halved because of antitrust rulings against the company and a general turndown in the NASDAQ.

15. Based on *Forbes'* 2004 list of the world's richest people. See www.forbes.com/lists/home.jhtml?passListId=10&passYear=2004&passListType=Person#searchlist.

16. 1900 data from Clyde Prestowitz, "Good but Not Good Enough," *World Link,* March/April 1994:31. Also read Lester Thurow, *The Future of Capitalism* (New York: Morrow, 1996):66. 2004 data from *Fortune 500.*

17. Based on *Forbes'* 2004 list of the world's richest people. Estimated fortunes are in parentheses and rounded in billions of dollars.

18. Data is based on 1995 rankings. *America Economia,* "Las 500 Mayores Empresas de America Latina," Edicion Annual 1996–97. The only three quasi-high-tech companies were Telmex, Telesp, and Telefonica.

19. One of the smartest observers of trends in Asia is Timothy Ong, editor of *Asia Inc.*

20. Robert H. Jackson, *Quasi-States: Sovereignty, International Relations, and the Third World* (London: Cambridge University Press, 1990). Francis Deng and Terrence Lyons, editors, *African Reckoning: A Quest for Good Governance* (Washington, D.C.: Brookings, 1998). Even organizations like CARE or the United Nations have pulled out of Angola and parts of Sudan because they consider the current situation hopeless.

21. Based on a speech by Lawrence Green, director of the CDC/WHO collaboration on tobacco and health. Reported in *Science* 290, November 17, 2000: 1291. And as health deteriorates and countries fall apart, there are ever fewer resources. In 1994, 400 doctors worked in Liberia's public hospitals, in 2000 only 85.

Chapter IV: Empires of the Mind

1. This book focuses on technology, not institution-building, although this is clearly a key development component. But remember that there are many countries with great institutions that end up economic disasters . . . because they lack technology as an engine of growth. If you want to remind yourself of just how crucial institutions and governance are, take a look at Bruce Scott's "The Great Divide in the Global Village," *Foreign Affairs,* Jan/Feb 2001.

2. Lee Kuan Yew provides a detailed road map in *From Third World to First: The Singapore Story 1965–2000* (Singapore: Singapore Press, 2000).

3. Bruce Scott of the Harvard Business School has done extensive research on the perils of being resource-rich. This table mostly uses data from the *World Development Report 2002* (World Bank Publications, 2001) and is complemented by data from the *New York Times Almanac 2005.* There is a constant debate as to how to measure per-capita wealth. You get different numbers when you do it on a purchasing-power scale or use other methods. But the overall lessons are the same in this case.

4. The transaction was part of the Treaty of Breda. Giles Milton has written a wonderful book on the spice trade and its dominance of global commerce, *Nathaniel's Nutmeg, or How One Man's Courage Changed the Course of History* (New York: Farrar Straus & Giroux, 1999).

5. Chart adapted from *The Economist,* April 17, 1999: 75.

6. Brunei may end up close to where it started. If you want a great historical novel on the area, check out *Kalimantaan* by C. S. Godshalk (New York: Henry Holt, 1998).

7. Want to tour the last millennium? See James Reston Jr., *The Last Apocalypse: Europe at the Year 1000 A.D.* (New York: Doubleday, 1998). Things did not get better over the next few centuries either: Read William Manchester's *A World Lit Only by Fire: The Medieval Mind and the Renaissance* (Boston: Little, Brown, 1992).

Chapter V: Data Drives Empires

1. Digital photography is simple, ever sharper, and easy to use-develop-modify. Contrast this with creating a Polaroid print, which requires more than 3,000 chemical reactions. (This is why Edwin Land, Polaroid's founder, garnered 535 patents, a record bested only by Thomas Alva Edison.) See Jessie Scanlon, "Flash Forward," *Wired,* January 2001.

2. Duncan J. Melville teaches a class in Mesopotamian and mathematics at St. Lawrence University in Canton, New York. (http://it.stlawu.edu/~dmelvill/mesomath/index.html). If you want to learn to count in cuneiform go to http://it.stlawu.edu/~dmelvill/mesomath/Numbers.html.

3. Nicolas Negroponte wrote a wonderful book on this: *Being Digital* (New York: Knopf, 1995). A more recent, more technical book is Frances Cairncross' *The Death of Distance: How the Communications Revolution Will Change Our Lives* (Boston: Harvard Business School Press, 1997); or see Neal Stephenson's *In the Beginning . . . Was the Command Line* (New York: Avon, 1999).

4. Various organizations monitor these trends, among them Gartner and Telegeography. For a summary see Juan Enriquez, "Latin America's Changing Media: Social, Political, and Economic Implications," Latin American Studies Association, September 24, 1998, Chicago.

5. Because the speed of change is so rapid, it is sometimes easier to follow the day-to-day changes in the digital world through magazines rather than books. I find *Wired, Fast Company,* and *Red Herring* useful ways to follow the intersection of technology and business.

6. Michael Lewis describes the uncertainty and fragility of big business beautifully in *The New New Thing: A Silicon Valley Story* (New York: W. W. Norton, 1999).

Chapter VI: Genetics . . . the Next Dominant Language

1. Richard Lewontin is a source of never-ending smart and wry observations: See *It Ain't Necessarily So: The Dream of the Human Genome and Other Illusions* (New York: NYRB, 2000).

2. Would you like to read the original paper or its English translation? See MendelWeb, edited by Roger B. Blumberg: www.netspace.org/MendelWeb/.

3. Stephen Jay Gould is a wonderful scientist, Red Sox fan, and teacher. Check out *The Mismeasure of Man* (New York: W. W. Norton, 1981). Also look at *Crossing Over: Where Art and Science Meet* (New York: Three Rivers Press, 2000), a collaboration between Gould and visual artist Rosamond Wolff Purcell.

4. The bane and gadfly of many scientists, Jeremy Rifkin, is a powerful speaker who has rallied a lot of people to his various causes. Check out *The Biotech Century: Harnessing the Gene and*

Remaking the World (New York: Jeremy P. Tarcher/Putnam, 1998). See also Greenpeace positions, www.greenpeace.org/~geneng.

5. Jared Diamond wrote a marvelous book on this subject, *Germs, Guns, and Steel: The Fates of Human Societies* (New York: W.W. Norton, 1997), in which he explains the rise of some civilizations as a function of access to food and domestic animals.

6. S. Tanksley and S. McCouch, "Seed Banks and Molecular Banks," *Science* 277, pp 1063–66.

7. Potatoes and tomatoes contain glycoalkaloids, which may cause spina bifida. Kidney beans contain phytohaemagglutinin, a poison. Peach seeds kill people with cyanogenic glycosides . . . If these foods were subject to the strict FDA requirements used for medicines, they would not be on supermarket shelves . . .

8. J. D. Watson and F. H. C. Crick, "A Structure for Deoxyribose Nucleic Acid," *Nature* 170 (April 2, 1953): 737. The article concludes with one of the great understatements in the history of science: "It has not escaped our notice that the specific pairing we have postulated immediately suggests a possible copying mechanism for the genetic material." You can read the original paper: http://biocrs.biomed.brown.edu/Books/Chapters/Ch%208/DH-Paper.html. You might also want to look at Watson's opinionated, short, and controversial *The Double Helix* (New York: Atheneum, 1968).

9. *Time* set up a fun page that profiles their picks for the top 100 scientists: www.time.com/time/time100/scientist/index.html.

10. And when you want to translate gene code into proteins, you have to be even more careful. Careless translations bring many pitfalls. Cantonese smile when they hear the words "Mercedes-Benz," because it can sound like "so clumsy, you die."

11. Not all changes in DNA lead to radically different outcomes. Through 1999, it was estimated that only 3 percent of the total DNA molecule actually coded—that is, provided specific instructions for life functions. These are the areas where genes are concentrated. However, the more we understand about the genome, the likelier it is that we will find vital functions within "noncoding" regions. Greg Verdine, one of Harvard's chemistry superstars, points out that all mammals have a modified version of cytosine (5-methylcytosine), which impacts gene expression so profoundly it could be considered a fifth nucleotide in the genetic language.

12. Chimps vary by about 1.5 percent. See A. G. Clark, "Chips for Chimps," *Nature Genetics* 16 (1999): 28–36. And M. Goodman, "The Genomic Record of Mankind's Evolutionary Roots," *American Journal of Human Genetics* 64 (1999): 31–39.

13. Henrik Kaessmann, Victor Wiebe, Svante Pääbo, "Extensive Nuclear DNA Sequence Diversity Among Chimpanzees," *Science* 286 (November 5, 1999): 1159–62.

14. One of the best overviews of how these variations will impact medicine is Samuel Broder and J. Craig Venter, "Sequencing the Entire Genomes of Free Living Organisms: The Foundation of Pharmacology in the New Millennium," *Annual Review of Pharmacology and Toxicology* 40 (2000): 97–102.

15. There are multiple studies on these types of genetic variations (BRCA-1, BRCA-2, P53). See, for example, www.oncolink.org/causeprevent/genetics/brca1/blank.html. Diverse reports in *The New England Journal of Medicine* (particularly January 16, 1996). Or J. M. Hall, M. K. Lee, B. Newman et al., "Linkage of Early-Onset Familial Breast Cancer to Chromosome 17q21," *Science* 250 (1990): 1684–89. Malkin, L.C. Strong et al., "Germ Line p53 Mutations in a Familial

Syndrome of Breast Cancer, Sarcomas, and Other Neoplasms," *Science* 250 (1990): 1233–38. Also, in November 2000, an Emory University scientist reported that a methyltransferase enzyme called DNMT-1 could play a key role in preventing the TMS-1 gene from instructing damaged or old cells to self-destruct. But note that only 5 to 7 percent of breast cancer cases are inherited. Folks like Paul Ewald argue there may also be a viral component. Recently, a similar pattern was found with AIDS. Those who have a RANTES gene are less susceptible to HIV and AIDS but have a high risk of serious asthma. *AIDS 2000*, vol. 14, no. 17; *Genes and Immunity*, December 2000, vol. 1, no. 8.

16. Eric Lander, director of MIT's Whitehead Institute, is one of the most engaging speakers and prolific researchers on genomics. He talked about this data during a conference at Harvard Medical School on November 9, 2000. He also explained that because man is such a new species (only about 7,000 generations old), there are only three or four common gene variations. For instance, Apolipoprotein E has E2, E3, E4. If you carry two copies of E4, you will probably get Alzheimer's. More details? See M. I. Kamboh, "Apolipoprotein E Polymorphism and Susceptibility to Alzheimer's Disease," *Human Biology* 67 (1995): 195–215. And A. D. Roses, "Apolipoprotein E Alleles as Risk Factors in Alzheimer's Disease," *Annual Review of Medicine*, 47 (1996): 387–400. The latest on this is Amanda Myers et al., "Susceptibility Locus for Alzheimer's Disease on Chromosome 10," *Science* (December 22, 2000): 2304. If you want something less technical, look at Rudolph Tanzi and Ann Parson's *Decoding Darkness: The Search for the Genetic Causes of Alzheimer's Disease* (Cambridge, Mass.: Perseus, 2000).

17. Knowledge of what makes you sick, how to cure it, or how to prevent it changes nearly every day. Various magazines cover the cutting-edge research, including *Science*, *Nature Medicine*, and *Nature Genetics*. On a daily basis, you can look at various sites like www. biospace.com/b2/news.cfm#biobeat.

18. J. C. Knight et al., "A Polymorphism That Affects OCT-1 Binding to the TNF Promoter Region Is Associated with Severe Malaria," *Nature Genetics* 22 (1999): 145–50. Are you a night owl? Louis J. Ptacek found the world's first circadian-rhythm gene; if you are, substitute glycine for serine at position 662 in the hPer2 protein—you will fall asleep at 7 P.M. and wake at 2 A.M. See www.hhmi.org/news/ptacek2.html.

19. Aaron Zitner wrote a short sketch on Haseltine: "An Ego in a Lab Coat Seeks the Genetic Fountain of Youth," *Los Angeles Times*, August 23, 2000. Just one of hundreds of articles on this controversial man. Take a look at this week's discoveries at www.humangenomesciences.com.

20. Actually, there is an interesting debate going on as to whether this is true. See Paul Ewald's *Plague Time: How Stealth Infections Cause Cancers, Heart Disease, and Other Deadly Ailments* (New York: Free Press, 2000).

21. Simon Melov et al., "Extension of Life-Span with Superoxide Dismutase/Catalase Mimetics," *Science* 289 (September 1, 2000): 1567–69. You might also want to look at "Extension of Cell Lifespan and Telomere Length in Animals Cloned from Senescent Somatic Cells," *Science*, April 28, 2000.

22. See www.hsph.harvard.edu/organizations/bdu/summary.html. DALYs are a measure of years of life lost due to premature mortality plus number of years lived with a disability, adjusted for the severity of the disability. The most complete study of these changes was done for the World Health Organization by Alan D. Lopez and Christopher Murray: *The Global*

Burden of Disease. Some technical terms in the chart are simplified; they should read: lower respiratory infections, conditions in perinatal period, unipolar major depression, ischemic heart disease, cerebrovascular disease, and chronic obstructive pulmonary disease.

Chapter VII: Genetics Is . . . a Hockey Stick

1. Photograph provided by Craig Venter. Craig Venter and Robert Fleishmann were also partners in crime, sequencing two chromosomes of the cholera pathogen, *Vibrio cholerae,* together with other scientists. *Nature* 406 (August 3, 2000).

2. Robert D. Fleishmann et al., "Whole Genome Random Sequencing and Assembly of *Haemophilus influenzae,*" *Science* 269 (July 28, 1995): 496–512. Venter co-founded an extraordinary think tank, The Institute for Genomic Research. His team includes Nobel laureate Hamilton Smith and one of the world's most talented biologists, Claire Fraser (who is also Venter's wife). See www.tigr.org.

3. Take a look at www.tigr.org/tdb/CMR/ghi/htmls/SplashPage.html.

4. If there is a particular organism or disease you are interested in, you can look at all complete microbial genomes as well as the original articles at www.tigr.org/tdb/mdb/mdbcomplete.html. There are also many more microbial genomes in progress: www.tigr.org/tdb/mdb/mdbin-progress.html.

5. Tom Standage wrote a fun book on this: *The Neptune File: A Story of Astronomical Rivalry and the Pioneers of Planet Hunting* (New York: Walker, 2000).

6. For example, Venter was able to identify thousands of new genes in a short period; see "Identification of New Human Receptor and Transporter Genes by High Throughput cDNA (EST) Sequencing," *Journal of Clinical Pharmacology* 45 (1993): 355–60.

7. *Nature* 377 (1995): 3–174. One can see a much more updated list, and of full sequences, at www.tigr.org. In GenBank: www.ncbi.nlm.nih.gov/ or, in a more general directory: http://nscp.snap.com/directory/category/0,16,nscp-33850,00.html?st.sn.sr.1.cat.

8. *Diamond* v. *Chakrabarty* (1980).

9. USPTO bulletin of October 23, 1996, "PTO Announces New Policy to Process Gene Sequence Biotechnology Patents." See also the debate that took place at UCSD's International Center on April 16, 1996. Or Eliot Marshall, "Companies Rush to Patent DNA," *Science* 275 (February 7, 1997): 780–81. A good general source for current issues and debates in this area is *The Journal of Biolaw and Business.* Many EST patents are likely to be challenged on the basis of 35 USC 101, which sets the utility guidelines for patents.

10. The Tasmanian tiger may take a while, because they are using DNA from an animal preserved in alcohol in 1866 but . . . in October 2000 a cow named Bessie was pregnant with a rare Asian ox, which had been cloned from a single skin cell of a dead male (the first endangered species to be cloned, and the first cloned animal to gestate in another species). Noah was born January 8, 2001, but unfortunately died two days later from a bacterial infection common to calves. Next targets include giant pandas, Sumatran tigers, and an extinct Spanish mountain goat (bucardo). The company behind this work is Advanced Cell Technologies of Worcester, Massachusetts. See Robert P. Lanza et al., "Cloning an Endangered Species *(Bos gaurus)* Using Interspecies Nuclear Transfer," *Cloning,* vol. 2 (2):79–90. Also "Cloning Noah's Ark" at www.scientificamerican.com/2000/1100issue/1100lanza.html.

11. You can read about the implications of this in the Kurzweil-Dertouzos debate in MIT's *Technology Review,* Jan/Feb 2001.

12. See www.ncbi.nlm.nih.gov/Genbank/genbankstats.html.

Chapter VIII: The Most Powerful Information System

1. Ironically, one of the drivers of Blue Gene was losing Celera's business. When Celera asked for bids on its computer system, it provided the raw *H. influenzae* gene sequence and timed assembly on different machines. It took Compaq seven hours vs. IBM's eighty-seven . . .

2. One of the best thinkers on these issues is Ray Kurzweil; see *The Age of Spiritual Machines: When Computers Exceed Human Intelligence* (New York: Viking, 1999). His views upset many, including the chief technology officer at Sun Microsystems, Bill Joy, who wrote an extraordinary rebuttal in the April 2000 *Wired* titled "Why the Future Does Not Need Us." www.wired.com/wired/archive/8.04/joy.html.

3. Take a look at what Church is working on: http://arep.med.harvard.edu/.

4. See David Malakoff, "Biocomputing," *Science,* June 11, 1999.

5. Hal Varian and Peter Lyman, "How Much Information?" at www.sims.berkeley.edu/how-much-info/print.html. Meanwhile, a series of Web companies are trying to catalog all global digital knowledge.

6. One of the leading authorities in the field is Nagoya University's Makoto Fujita.

7. You can read a great description of Clark and Silicon Valley in Michael Lewis' *The New New Thing.*

8. There is a really interesting article in *Science* that outlines the initial possibilities. James R. Heath et al., "A Defect Tolerant Computer Architecture: Opportunities for Nanotechnology," 280 (June 12, 1998): 1716.

9. www.beowulf.org.

10. You can see what is discovered daily by looking at the Protein Data Bank: www.rcsb.org/pdb/index.html.

11. Ironically, it was a gift from the reverend, a Penaud toy helicopter, that began the kids' fascination with flying machines. If you would like to see an exact replica of their wondrous plane fly, show up at Kitty Hawk, North Carolina, on December 17, 2003, at 10:35 A.M. for the centennial commemorative flight. Bishop Wright's encounter with technology was narrated by clergyman Bruce Larson and told to me by one of Malaysia's living national treasures, Datuk Paddy Bowie.

12. University of Michigan's Wendell Weber has done a lot of work on how the same drug can help or hurt people and how to personalize prescriptions (pharmacogenomics). If you have a faulty enzyme in your body like CYP2D6 PM, then you may be poisoned by thirty drugs . . . CYP2D6 URMS means that thirty drugs will have no therapeutic effect . . . This is part of the reason why some treatments for cancer, multiple sclerosis, and asthma are useless or end up hurting the patient. These studies also have important implications for addictive disorders: aldehyde dehydrogeniks cannot tolerate alcohol; those with CYP2A6 PM cannot metabolize nicotine . . . Other CYP2D6 polymorphisms can make you immune to codeine's effects or so sensitive that it creates a rapid overdose.

13. The Federal Reserve Bank of Dallas has created an entertaining index comparing all kinds of prices across time. See the 1997 Annual Report, "Time Well Spent: The Declining Real Cost of Living in America": www.dallasfed.org/htm/pubs/annual/arpt97.html.

14. Michael Eisen, who works at the University of California Berkeley and Lawrence Berkeley National Lab, found that "genes that are co-expressed tend to share function." He discovered this by using microarrays to observe how yeast genes react to a series of stimuli like stress, heat, cold, starvation, beer. This provides vital clues for the diagnosis and treatment of cancer. For instance, almost all patients with DLBCL, the most common form of non-Hodgkin's lymphoma, react well to initial treatments. But only half will end up living a long time, while the others die quickly. By looking at patterns of gene expression, M.D.s can now figure out what the likely outcome of treatment will be and whether to follow a standard prescription (for germinal cell B type) or a far more aggressive treatment (if active B type). This technique is also being used to refine diagnosis for breast cancer. If a microarray shows that you are keratin 5 or 17 negative you have a good chance of surviving, but if you express either or both, you are in trouble.

15. Prof. Doug Brutlag, who teaches medicine and biochemistry at Stanford, gave a good overview of the problem during the AAAS meetings in San Francisco, February 17, 2001.

Chapter IX: Nano World

1. And you could look at it on an atom-by-atom scale while figuring out what each atom is made of. Chad A. Mirkin, Seunghun Hong, and Jin Zhu, "Multiple Ink Nanolithography: Toward a Multiple-Pen Nano-Plotter," *Science,* 286 (October 15, 1999): 523–25.

2. One of the great minds of the twentieth century, Richard Feynman, predicted this four decades ago. You can read his classic speech to the American Physical Society December 29, 1959, at www.zyvex.com/nanotech/feynman.html. You might also enjoy his book *Surely You're Joking, Mr. Feynman* (New York: W.W. Norton, 1985).

3. There are alternative ways to power nano machines that may turn out to be easier to replicate, such as exothermic chemical reactions. See A. K. Schmid, N. C. Bartelt, and R. Q. Hwang, "Alloying at Surfaces by the Migration of Reactive Two-Dimensional Islands," *Science* 290 (November 24, 2000): 1561–74.

4. Hugh Aldersey-Williams gives a good overview of these discoveries in *The Most Beautiful Molecule: The Discovery of the Buckyball* (New York: John Wiley & Sons, 1995).

5. Cess Dekker's lab at the Delft University of Technology made the first nano transistor in 1998.

6. The first such creatures were reported in *Nature* on August 31, 2000. For an update on current work, look at Brandeis University's DEMO lab (particularly the research by Jordan Pollack and Hod Lipson) and at MIT's artificial-intelligence lab.

7. Ricky K. Soong et al., "Powering an Inorganic Nanodevice with a Molecular Rotor," *Science,* November 24, 2000. Want more? Take a look at two Web sites: the Cornell Nanotechnology Center, www.nbtc.cornell.edu, and the Cornell Nanofabrication Facility, www.cnf.cornell.edu.

8. Physicists, chemists, biologists, and computer scientists have been working together at U.N.C. Chapel Hill to build nanomanipulators. See a special *Science* issue on nanotechnology:

November 24, 2000. (It is interesting to compare this issue with the journal's November 21, 1990, issue on nanotech, "Engineering a Small World.")

9. Carl Woese was the first to argue for a third branch of life based on this bacterium, but few believed this until after TIGR published the genomic sequence.

10. Bult et al., *Science* 273 (August 23, 1996): 1058–73. See also "Genome Data Shakes Tree of Life," *Science* 280 (May 1, 1998): 672–74.

11. Edward DeLong, "Archaeal Means and Extremes," *Science* (April 24, 1998): 542–43.

12. Owen White et al., "Genome Sequence of the Radioresistant Bacterium *Deinococcus radiodurans*," *Science* 286 (November 19, 1999): 1571–77. The bacterium survives not only radiation but also starvation, desiccation, heat . . . It does so by having many copies of the same genes and by being very good at repairing damage. (It can survive 3 Mrads.) The three chromosomes can be blown apart and then stitch themselves back together within 24 hours. For greater detail, see Paula A. Kiberstis, "Robust Survival Strategy?" *Genes Development* 14 (2000): 777.

13. R. John Parkes, "A Case of Bacterial Immortality?" and Russell H. Vreeland, William Rosenzweig, and Dennis W. Powers, "Isolation of a 250-Million-Year-Old Halotolerant Bacterium from a Primary Salt Crystal," *Nature*, October 19, 2000. There is also presumed evidence of life in a massive lake that sits under Antarctica. These life forms were isolated long before humans walked the planet and have never been seen: Erica B. Goldman, "A Tale of Two Lakes," *The Sciences*, Jan/Feb 2001.

14. Louis Allamandola of NASA's Ames Research Center concluded: "This discovery implies that life could be everywhere in the universe."

15. The basic experiment was "Global Transposon Mutagenesis and a Minimal Mycoplasma Genome," by Clyde A. Hutchison III et al., *Science* 286 (December 10, 1999): 2165–69. The ethics panel published Mildred K. Cho et al., "Ethical Considerations in Synthesizing a Minimal Genome," *Science* 286 (December 10, 1999): 2087–90.

16. If you would like to stay up to date on nano research, take a look at the World Technology Evaluation Center's Report, "R&D Status and Trends in Nanoparticles, Nanostructured Materials, and Nanodevices in the United States": http://itri.loyola.edu/nano/us_r_n_d/toc.htm.

17. You can see a picture of Alba at www.ekac.org. Kac's concerns are part of a broader movement. Mass MoCA did a show called "Unnatural Science" in spring 2000, and Exit Art, a Soho, N.Y., gallery, did "Paradise Now: Picturing the Genetic Revolution" in September 2000. In January 2001, New York's International Center of Photography exhibited "Perfecting Mankind: Eugenics and Photography."

Chapter X: Revolution . . . in a Few ZIP Codes

1. USPTO, "TAF Special Report: All Patents, All Types," March 2003. The problem is not just within the U.S. market. In the United Kingdom, for instance, patents granted grew 16 percent between 1997 and 1998; during that period, Korea received 430 patents, Taiwan 180, and Argentina, Brazil, Mexico, and Venezuela—together—received thirteen.

2. *Mexico Social 1996–98,* Banamex-Accival, 1998, 402, 403, 661.

3. You can look at these trends in the aggregate or country by country in the World Bank's annual *World Development Report.*

4. In 1982, the Brazilian patent office granted 8,808 patent and trademark requests; in 1996, it granted 2,501. From 1993 onward, there was an increase in petitions for patent protection, but not in patents granted. See www.inpi.gov.br.

5. www.impi.gob.mx/anexo97a.htm.

6. Indicadores de Actividades Científicas y Tecnológicas, 1997 and 2003. Ironically, the Mexican outfit with the most patents granted was a part not of the private sector, but of government. The Mexican Petroleum Institute was granted fifteen patents.

7. Alcorta Ludovico and Wilson Peres, "Innovation Systems and Technological Specialization in Latin America and the Caribbean," *Research Policy* 26, 1998. Their index of technological specialization measures the ratio of high- and medium-tech products exported to Organization for Economic Cooperation and Development (OECD) countries versus low-tech products. In 1987, China, Indonesia, Malaysia, and Thailand were far below Latin America. By 1994, they were on a par or above. "Asian tigers" includes Korea, Singapore, Hong Kong, and Taiwan.

8. Lester Thurow has written on this phenomenon in *Building WEALTH: The New Rules for Individuals, Companies, and Nations in a Knowledge-Based Economy* (New York: HarperCollins, 1999).

9. For general patent numbers, look at www.uspto.gov/web/offices/ac/ido/oeip/taf/tafp.html; issue dates can be found at www.uspto.gov/web/offices/ac/ido/oeip/taf/h_counts.htm.

10. Patents had been issued previously. On March 6, 1646, Joseph James received the first American mechanical patent to protect the manufacturing of scythes. The general court of Massachusetts issued it.

11. Professor Jeff Madrick has done some interesting work on these trends. He found U.S. workers of median age 35–44 earned 9 percent less than those working 25 years earlier and that incomes for those 45–54 have stagnated since 1973.

12. The literature on why crime has dropped in the United States over the past decade and whether this trend will continue is endless. You can get a sense of the debate from Laura Helmuth, "Has America's Tide of Violence Receded for Good?" in *Science,* July 28, 2000.

13. Rodger Doyle, "Coke, Crack, Pot, Speed, et al.," *Scientific American,* January 2001.

14. "Doctoral Scientists and Engineers in the United States, 1995 Profile," National Academy of Sciences: 5. There are some indicators that things may be getting slightly better; NAS files from September 1999 show that 3.1 percent of engineering Ph.D.s were black and 2.8 percent were Hispanic. But the gaps are huge; in 1999, 57 percent of white seventeen-year-olds were able to analyze scientific data and procedures . . . 27 percent of Hispanics . . . 12 percent of blacks. See http://nces.ed.gov/pubsearch/pubsinfo.asp?pubid=2001034, Table 129: percentage of students at or above selected science proficiency levels.

15. *LASA 2000: Hands Across the Hemisphere in the New Millennium.* Miami, March 16–18, 2000.

16. Figures come from Department of Education, National Center for Education Statistics, "High School Drop Outs by Race-Ethnicity and Recency of Migration," June 2000. They cover high school students in 1997.

17. Marcelo and Carola Suarez Orozco head Harvard's immigration project and have written about the implications of these trends. See *Crossings: Mexican Immigration in Interdisciplinary*

Perspectives (DRCLAS, Harvard University Press, 1999). Also look at the work of U.C. San Diego scholar Wayne Cornelius.

18. One place you can follow these trends is www.aflcio.org/paywatch/ceopay.htm.

19. This time period is from initial commercial application, not from invention.

20. Of course, lawyers and politicians were involved, so it actually took a year to make what happened de facto become de jure. Read Charles S. Maier's *Dissolution: The Crisis of Communism and the End of East Germany* (Princeton: Princeton University Press, 1997).

Chapter XI: Technology Is Not Kind . . . It Does Not Say "Please"

1. Juan Enriquez, "Genomics and the World's Economy," *Science,* August 14, 1998.

2. Ray Goldberg, *A Concept of Agribusiness.* He has written or edited a book a year on the theme. See, for example, *Agribusiness Coordination* (Cambridge: Harvard University Press, 1968).

3. Ray Goldberg helped guide Pioneer for years as one of its most active directors. If you want to look at these trends in greater depth, see Juan Enriquez and Ray Goldberg, "Transforming Life, Transforming Business: The Life Science Revolution," *Harvard Business Review,* March/April 2000. Or Juan Enriquez, *Gene Research, the Mapping of Life, and the Global Economy,* Harvard Business School Case 599-016.

4. The trends are clear in Margaret F. Brennan, Carl E. Pray, and Ann Courtmanche, "Impact of Industry Concentration on Innovation in the U.S. Plant Biotech Industry," Rutgers University Dept. of Agricultural, Food, and Resource Economics (Draft June 23, 1999).

5. A Cornell team led by Charles Arntzen has created a Norwalk virus vaccine in potatoes and is working on bananas. See the July 2000 issue of the *Journal of Infectious Diseases* or, if you want to be more up to date, http://bti.cornell.edu/index.html.

6. Glennerster and Kremer, "The Need for Vaccine Research," *Brookings Policy Brief,* May 2000.

7. Dean DellaPenna, "Nutritional Genomics: Manipulating Plant Micronutrients to Improve Human Health," *Science* 285 (1999): 375–79.

8. A. W. S. Chan, K. Y. Chong, C. Martinovich, C. Simerly, and G. Schatten, "Transgenic Monkeys Produced by Retroviral Gene Transfer into Mature Oocytes," *Science* 291 (2001): 309–12.

9. There is a lot written on the ethical, legal, and social implications of gene research. See www.niehs.nih.gov/envgenom/elsi.htm.

10. Let's talk about sex . . . Gina Bari Kolata, *Clone: The Road to Dolly, and the Path Ahead* (New York: William Morrow, 1998). Lori B. Andrews, *The Clone Age: Adventures in the New World of Reproductive Technology* (New York: Henry Holt, 1999).

11. Juan Enriquez, "Green Biotech and European Competitiveness," *Trends in Biotechnology,* April 2001.

12. This process of creative destruction was identified over half a century ago by an Austrian economist turned Harvard professor, Joseph Schumpeter. See Joseph Schumpeter, *Business Cycles: A Theoretical, Historical and Statistical Analysis of the Capitalist Process* (New York: McGraw-Hill, 1939).

13. Charles J. Whalen, "Today's Hottest Economist Died Fifty Years Ago," *Business Week,* December 11, 2000.

Chapter XII: Sleepless . . . (and Angry) in Seattle

1. Daniel Hecker, "High Technology Employment: A Broader View," *Monthly Labor Review,* June 1999.

2. Latin for "Let the buyer beware."

3. Based on Yahoo! Finance, Hoover's Company Capsules, and Excite as well as annual reports for companies. Net income is income available to common stock for trailing twelve months. Data is from February 2001.

4. Lou Gerstner quoted in *Technology Review,* February 2001. This publication, which describes itself as "MIT's magazine of innovation," is a good way to follow some tech trends.

Chapter XIII: High Tech . . . High Pay . . . High Mobility . . .

1. By Berjaya, MUI, and Kuala Lumpur Kempong, respectively. Michael Backman covers these trends in *Asian Eclipse: Exposing the Dark Side of Business in Asia* (New York: Wiley, 1999).

2. Daniel Pink started a widespread and controversial debate with "Free Agent Nation" in *Fast Company,* January 1998.

3. These figures cover real earnings by males (U.S. Department of Labor). Calculated by *The Economist.* You might enjoy the magazine's "Survey: The New Economy," September 23, 2000. Rice University's Stephen L. Klinkenberg has done similar work and found a similar correlation between education and income. There are also several OECD studies detailing the importance of education and providing indicators by member states.

4. U.S. Occupational Employment Statistics Survey. These trends hold for recent college grads. A chemical-engineering major (1996–97) had a starting salary of $42,758, a computer scientist $36,964, a mathematician $32,055, a business administrator $28,506, a social scientist $24,232, a liberal-arts major $24,081, and a journalist $22,102; see Patrick Scheetz, *Recruiting Trends 1996–97* (Lansing: Michigan State University, 1998). You can look at other professions at www.usnews.com/usnews/edu/beyond/bcsalary.htm.

5. www.moe.go.kr/english/edukorea/edukorea2/edukorea2.html.

6. Thanks to Professors Byung Kook Kim and Young-Jin Kim for their time and ideas. Minister Dal Ho Chung arranged an extraordinary visit to his country.

7. See *Digest of Education Statistics 1998.* Particularly Chapter 6, International Comparisons of Education. Math winners . . . comes from International Association for the Evaluation of Educational Achievement, Mathematics and Science Achievement in Secondary School: IEA's Third International Mathematics and Science Study, 1997, by Albert E. Beaton et al. See http://nces.ed.gov/pubs99/digest98/chapter6.html. The fourth-grade class that did OK was then retested as eighth graders. They too failed to perform.

8. The conviction was appealed and thrown out on a technicality, but it was only in 1968 that the Supreme Court vindicated Scopes' arguments.

9. The board of education reversed itself in February 2001, two days after the human genome was published on the Web.

10. Ed Regis, "Zip Drive," *Wired,* January 2001. Or take a look at the project itself: NASA Technical Paper 3539 "Rapid Mars Transits with Exhaust Modulated Plasma Propulsion."

11. Dennis Taylor and Erik Espe, "Immigrants Impacting Valley Economy," *Silicon Valley/San Jose Business Journal,* July 16, 1999. Based on AnnaLee Saxenian's research.

12. Take a look at the current horror at the Commerce Department: www.bea.doc.gov/. By March 2000, Robert Samuelson pointed out (based on IMF statistics) that foreigners owned $1.4 trillion worth of U.S. stocks (7 percent of total), close to $900 billion of corporate bonds (a fifth of total), over one-third of publicly held federal debt ($1.3 trillion), and close to $1.2 trillion in direct investments (5 to 8 percent).

13. *The Economist* ran a survey of these trends: "Debt in Japan and America," January 22, 2000.

14. Total household debt increased from 85 percent of personal income (1992) to 103 percent in 1999.

15. "Catch-22" refers to a novel by Joseph Heller. Its title reflects a choice where you are damned if you do . . . and damned if you don't.

16. Data based on a study by Teresa Bracho, Colegio de Mexico. The country's upper middle class is defined as those whose income is in the ninth decile—that is, those immediately below the top 10 percent of income earners.

17. Professor Teresa Bracho, Rojitas' mentor, uses this story as an example of the problems within Mexico's educational system.

18. Templo Mayor, *Reforma,* June 8, 2000. Science refers to CONACYT budget, foodstuffs to CONASUPO.

19. Digest of Education Statistics 2000: http://nces.ed.gov/pubsearch/pubsinfo.asp?pubid=2001034.

20. www.imf.org/external/pubs/ft/wp/2000/wp0044.pdf.

Chapter XIV: The Digital-Genomics Diaspora

1. The reasons for this massive migration include natural disasters, wars, or government repression. But there is also evidence of economic migration to improve one's lifestyle, particularly among the educated. Take a look at the United Nations' *World Migration Report 2000* and at a map of H1-B visa applications in the United States.

2. Take a look at the MIT study "Retreat of the Firm and the Rise of Guilds: The Employment Relationship in the Age of Virtual Business": http://cdiatwork.com/. You can also look at these trends and recent surveys at www.telecommute.org/. About 9.9 percent of the U.S. workforce is self-employed, and many corporations allow their wired employees to "WAH" (work at home).

3. Keith Pavitt, "Academic Research in Europe," Science and Technology Policy Research Paper 43, University of Sussex. D. M. Dalton and P. Yoshida, *Globalizing Industrial Research and Development,* U.S. Department of Commerce, 1999.

4. You can get a detailed breakdown of what students intend to do by country from *Statistical Profiles of Foreign Doctoral Recipients in Science and Engineering: Plans to Stay in the United States,* Jean M. Johnson, NSF 99-304. Some stay just for a postdoctoral program, but many end up in permanent jobs. See also National Academy of Sciences. "International Benchmarking of U.S. Mathematics Research" (prepublication copy, 1997), p. 40. Based on data from the NSB 1996.

5. There are actually more countries than this, but I am considering only United Nations members. If you are curious who became what when, take a look at www.un.org/members/.

6. For a more detailed discussion of these trends, see Juan Enriquez, "Too Many Flags?" in *Foreign Policy,* fall 1999, which is part of a forthcoming book, *Flags, Borders, Anthems, and Other Myths: The Impulse Toward Secession and the Americas.*

7. Curious? Read Katie Hafner and Matthew Lyon's *Where Wizards Stay Up Late: The Origins of the Internet.* (New York: Simon & Schuster, 1996).

8. Professor AnnaLee Saxenian wrote *Silicon Valley's New Immigrant Entrepreneurs* (Public Policy Institute of California, 1999) as well as a breakthrough book, *Regional Advantage: Culture and Competitiveness in Silicon Valley and Route 128* (Cambridge: Harvard University Press, 1994) . . . Please read it—it's great.

9. Commodities allow rebels to finance their mayhem without producing anything. Past conflict increases your chance of further war 40 percent, with the probability falling 1 percent for each year of peace. But if there is a substantial population abroad, in rich countries, conflict is more likely. A large American population increases the possibility of conflict 36 percent; a small diaspora reduces it to 6 percent. Having 55 percent of youths in secondary school instead of 45 percent cuts the probability of conflict from 14 to 10 percent. Ethnically diverse countries have a 3 percent probability of conflict, whereas ethnic and religious homogeneity increase the chances of conflict to 23 percent. Paul Collier and Anke Hoeffler wrote a detailed paper on these trends and their implications: "Greed and Grievance in Civil War," World Bank Policy Research Working Paper 2355. Also: Paul Collier, "Economic Causes of Civil Conflict and their Implications for Policy," www.worldbank.org/research/conflict/.

Chapter XV: Time Warp

1. This chapter was inspired by Brian Alexander's (You) 2, *Wired,* February 2001. He does a great job of covering the current science and hype behind human cloning. See also Margaret Talbot, "A Desire to Duplicate," *New York Times Magazine,* February 4, 2001. There are a variety of Web sites including www.humancloning.org. Scientists are tiptoeing ever closer to human cloning; Jan Tesarik et al., "Chemically and Mechanically Induced Membrane Fusion: Non-Activating Methods for Nuclear Transfer in Mature Human Oocytes," *Human Reproduction* 15 (May 2000): 1149–54.

2. James D. Watson, "Moving Toward the Clonal Man: Is This What We Want?" *Atlantic Monthly,* May 1971, pages 50–53. He argued that legislation to stop cloning, and perhaps IVF, should be stopped globally before it was too late.

3. Two people were instrumental in enabling this revolution, because they conceived and built the world's leading sequencing machines: Leroy Hood and Mike Hunkapiller.

4. Many are not shy about where they are heading and what they hope to accomplish nor is their language subtle: A group of Nobel laureates et al. recently signed a declaration defending cloning of animals and perhaps humans by declaring: "Historically, the Luddite option, which seeks to turn back the clock and limit or prohibit the application of already existing tech-

nologies, has never proven realistic or productive. The potential benefits of cloning may be so immense that it would be a tragedy if ancient theological scruples should lead to a Luddite rejection of cloning." See "Declaration in Defense of Cloning and the Integrity of Scientific Research": www.secularhumanism.org/library/fi/cloning_declaration_17_3.html.

5. Christianity also allows occasional resurrection—i.e., Jesus and Lazarus—but it is not a constant, recurring phenomenon applicable to all believers.

6. And if you contact Nightlight Christian Adoptions in Fullerton, California . . . they may help you adopt a frozen embryo.

7. The lamb was cloned from a mammary cell . . . hence the name.

8. The preceding examples are based on a much more complete discussion of these issues carried out by a Renaissance man at Princeton who teaches molecular biology, ecology, and evolutionary biology. See Lee M. Silver, *Remaking Eden: Cloning and Beyond in a Brave New World* (New York: Avon, 1997).

9. David Cooper and Robert P. Lanza, *Xeno: The Promise of Transplanting Animal Organs into Humans* (Oxford: Oxford University Press, 2000). If you want to look at more technical papers: Irina A. Polejaeva et al., "Cloned Pigs Produced by Nuclear Transfer from Adult Somatic Cells," and Jeffrey L. Platt, "Xenotransplantation: New Risks, New Gains," *Nature* 407 (2000).

10. These questions were widely debated during the Swiss referendum on bioengineering in 1998. If you want to delve into the ethics of the debate, see Alex Mauron, "Is the Genome the Secular Equivalent of the Soul," *Science* 291 (February 2, 2001): 831–32. There are also religious questions . . . Eating a pig is un-kosher, but apparently some scholars feel transplanting parts of a pig is OK . . .

11. This accuracy data does not reflect the whole of the genome because the highly repetitive central section of chromosomes (centromeres) is not included. Nevertheless, it is an outstanding achievement. Few multiauthor science papers deserve a complete citation . . . But this paper does: J. Craig Venter, Mark D. Adams, Eugene W. Myers, Peter W. Li, Richard J. Mural, Granger G. Sutton, Hamilton O. Smith, Mark Yandell, Cheryl A. Evans, Robert A. Holt, Jeannine D. Gocayne, Peter Amanatides, Richard M. Ballew, Daniel H. Huson, Jennifer Russo Wortman, Qing Zhang, Chinnappa D. Kodira, Xiangqun H. Zheng, Lin Chen, Marian Skupski, Gangadharan Subramanian, Paul D. Thomas, Jinghui Zhang, George L. Gabor Miklos, Catherine Nelson, Samuel Broder, Andrew G. Clark, Joe Nadeau, Victor A. McKusick, Norton Zinder, Arnold J. Levine, Richard J. Roberts, Mel Simon, Carolyn Slayman, Michael Hunkapiller, Randall Bolanos, Arthur Delcher, Ian Dew, Daniel Fasulo, Michael Flanigan, Liliana Florea, Aaron Halpern, Sridhar Hannenhalli, Saul Kravitz, Samuel Levy, Clark Mobarry, Knut Reinert, Karin Remington, Jane Abu-Threideh, Ellen Beasley, Kendra Biddick, Vivien Bonazzi, Rhonda Brandon, Michele Cargill, Ishwar Chandramouliswaran, Rosane Charlab, Kabir Chaturvedi, Zuoming Deng, Valentina Di Francesco, Patrick Dunn, Karen Eilbeck, Carlos Evangelista, Andrei E. Gabrielian, Weiniu Gan, Wangmao Ge, Fangcheng Gong, Zhiping Gu, Ping Guan, Thomas J. Heiman, Maureen E. Higgins, Rui-Ru Ji, Zhaoxi Ke, Karen A. Ketchum, Zhongwu Lai, Yiding Lei, Zhenya Li, Jiayin Li, Yong Liang, Xiaoying Lin, Fu Lu, Gennady V. Merkulov, Natalia Milshina, Helen M. Moore, Ashwinikumar K. Naik, Vaibhav A. Narayan, Beena Neelam, Deborah Nusskern, Douglas B. Rusch, Steven Salzberg, Wei Shao, Bixiong Shue, Jingtao Sun, Zhen Yuan Wang, Aihui Wang, Xin Wang, Jian Wang, Ming-Hui Wei, Ron Wides, Chunlin Xiao, Chunhua Yan, Alison Yao, Jane Ye, Ming Zhan, Weiqing Zhang,

Hongyu Zhang, Qi Zhao, Liansheng Zhengl Fei Zhong, Wenyan Zhong, Shiaoping C. Zhu, Shaying Zhao, Dennis Gilbert, Suzanna Baumhueter, Gene Spier, Christine Carter, Anibal Cravchik, Trevor Woodage, Feroze Ali, Huijin An, Aderonke Awe, Danita Baldwin, Holly Baden, Mary Barnstead, Ian Barrow, Karen Beeson, Dana Busam, Amy Carver, Angela Center, Ming Lai Cheng, Liz Curry, Steve Danaher, Lionel Davenport, Raymond Desilets, Susanne Dietz, Kristina Dodson, Lisa Doup, Steven Ferriera, Neha Garg, Andres Gluecksmann, Brit Hart, Jason Haynes, Charles Haynes, Cheryl Heiner, Suzanne Hladun, Damon Hostin, Jarrett Houck, Timothy Howland, Chinyere Ibegwam, Jeffery Johnson, Francis Kalush, Lesley Kline, Shashi Koduru, Amy Love, Felecia Mann, David May, Steven McCawley, Tina McIntosh, Ivy McMullen, Mee Moy, Linda Moy, Brian Murphy, Keith Nelson, Cynthia Pfannkoch, Eric Pratts, Vinita Puri, Hina Qureshi, Matthew Reardon, Robert Rodriguez, Yu-Hui Rogers, Deanna Romblad, Bob Ruhfel, Richard Scott, Cynthia Sitter, Michelle Smallwood, Erin Stewart, Renee Strong, Ellen Suh, Reginald Thomas, Ni Ni Tint, Sukyee Tse, Claire Vech, Gary Wang, Jeremy Wetter, Sherita Williams, Monica Williams, Sandra Windsor, Emily Winn-Deen, Keriellen Wolfe, Jayshree Zaveri, Karena Zaveri, Josep F. Abril, Roderic Guigó, Michael J. Campbell, Kimmen V. Sjolander, Brian Karlak, Anish Kejariwal, Huaiyu Mi, Betty Lazareva, Thomas Hatton, Apurva Narechania, Karen Diemer, Anushya Muruganujan, Nan Guo, Shinji Sato, Vineet Bafna, Sorin Istrail, Ross Lippert, Russell Schwartz, Brian Walenz, Shibu Yooseph, David Allen, Anand Basu, James Baxendale, Louis Blick, Marcelo Caminha, John Carnes-Stine, Parris Caulk, Yen-Hui Chiang, My Coyne, Carl Dahlke, Anne Deslattes Mays, Maria Dombroski, Michael Donnelly, Dale Ely, Shiva Esparham, Carl Fosler, Harold Gire, Stephen Glanowski, Kenneth Glasser, Anna Glodek, Mark Gorokhov, Ken Graham, Barry Gropman, Michael Harris, Jeremy Heil, Scott Henderson, Jeffrey Hoover, Donald Jennings, Catherine Jordan, James Jordan, John Kasha, Leonid Kagan, Cheryl Kraft, Alexander Levitsky, Mark Lewis, Xiangjun Liu, John Lopez, Daniel Ma, William Majoros, Joe McDaniel, Sean Murphy, Matthew Newman, Trung Nguyen, Ngoc Nguyen, Marc Nodell, Sue Pan, Jim Peck, Marshall Peterson, William Rowe, Robert Sanders, John Scott, Michael Simpson, Thomas Smith, Arlan Sprague, Timothy Stockwell, Russell Turner, Eli Venter, Mei Wang, Meiyuan Wen, David Wu, Mitchell Wu, Ashley Xia, Ali Zandieh, Xiaohong Zhul, "The Sequence of the Human Genome," *Science* 291 (February 16, 2001): 1304–51.

The International Human Genome Sequencing Consortium also matched this extraordinary achievement with a different version of the genome published the same week: see *Nature* 409 (2001) . . . Got a lot of time on your hands and want to play *Where's Waldo?* Take a look at the hundreds of photographs that form the double helix on the cover of *Nature* and find the one of Watson and Crick.

12. Reported in *The New York Times* based on data from Dr. Mani Subramanian from Celera. Chart does not include 5.1 percent miscellaneous genes.

13. The experiments occurred during 1998 and 1999 but were not reported until February 2001 out of fear that rogue states might use the data to create a vaccine-resistant smallpox. Ronald J. Jackson et al., "Expression of Mouse Interleukin-4 by a Recombinant Ectromelia Virus Suppresses Cytolytic Lymphocyte Responses and Overcomes Genetic Resistance to Mousepox," *Journal of Virology* 75 (February 2001): 1205–10. These threats are likely to grow. Countries like the Soviet Union ignored treaty obligations and built massive biowarfare capabilities. See, for example, Ken Alibek, *Biohazard: The Chilling True Story of the Largest Covert*

Biological Weapons Program in the World—Told from the Inside by the Man Who Ran It (New York: Random House, 1999). If you want to look at what policymakers are thinking, see Joshua Lederberg, ed., *Biological Weapons: Limiting the Threat* (Cambridge: MIT Press, 1999).

14. William Manchester, *A World Lit Only by Fire* (Boston: Little Brown, 1992).

15. He first worked for a pharmaceutical company called IVAX.

16. J. M. Roberts, *History of the World* (Oxford: Oxford University Press, 1993). One of Asia's great minds, Timothy Ong, has explored the implications of these conclusions for the current economic crisis.

C R E D I T S

PG #	DESCRIPTION	CREDIT LINE
16	Angkor Wat	Keith Gunnar/National Audubon Society/Photo Researchers
26	Circuit boards	Astrid & Hanns-Frieder Michler/Science Photo Library/Photo Researchers
35	ENIAC	Jerry Cooke/CORBIS
57	Sultan of Brunei	AFP/CORBIS
59	Churchill	CORBIS
62	Lascaux	Francis G. Mayer/CORBIS
63	Cuneiform	Photo Researchers
69	Case and Levin	Reuters NewMedia Inc./CORBIS
73	Mendel	Bettmann/CORBIS
77	Crick and Watson	A. Barrington Brown/Science Photo Library/Photo Researchers
89	Craig Venter	Photograph by Bill Geiger/Courtesy of Craig Venter
93	Dr. Leder	Courtesy of Phillip Leder, M.D., Harvard Medical School, Department of Genetics
97	Cloned pigs	Reuters NewMedia Inc./CORBIS
104	Kasparov	Najlah Feanny/Stock Boston/Picturequest
129	Buck Fuller	Hans Namuth/Photo Researchers
152	Factory workers	Annie Griffiths Belt/CORBIS
162	Berlin Wall	AFP/CORBIS
169	Bonobos	John Giustina/BRUCE COLEMAN INC.

I N D E X

Blue Gene supercomputer (IBM), 105, 129
Boehringer family, 42
bonobos, as medical test animals, 81
Boston/Cambridge, Massachusetts, 101, 158.
　　See also Harvard University
Brazil
　　cloning in, 211
　　disparities in, 157, 161
　　exports of, 151
　　and 427:1 gap, 23, 24
　　fragmentation in, 204
　　knowledge creation in, 139, 140, 142, 144,
　　　146, 151, 152
　　natural resources in, 49, 53
　　politics in, 157
　　as superpower, 224
　　wealth in, 30, 39, 49, 53, 161
BRCA-1, 82
Broder, Sam, 222
Brown, Louise Joy, 214, 215
Brunei, 57, 198
Bubka, Sergei, 49
Buenos Aires stock exchange, 141
Buffett, Warren, 40
Burma. *See* Myanmar
business
　　restructuring of U.S., 50
　　See also companies/corporations

C

California, 158, 160, 211
Campbell's, 9
Canada
　　disparities in, 157
　　economic growth in, 155
　　education in, 189
　　exports of, 151, 152, 203
　　and foreign students in U.S., 203
　　knowledge creation in, 142, 151, 152, 157
　　Microsoft compared with, 172
　　natural resources in, 53
　　trade with, 155
　　wealth in, 53, 157
cancer, 82, 83, 123, 222
Canon, 154
Caplan, Art, 225
carbon, and nano world, 128–29, 130–31
Carnegie Mellon University, 186

Celera, 92, 99, 108, 109–11, 125, 178, 221
Chang-Diaz, Franklin Ramon, 191
change
　　adopting and adapting to, 224
　　of codes, 3–6
　　and convergence, 134–36
　　and "creative destruction," 172
　　and development of digital language, 70
　　fear of, 175–83
　　in genome, 75–77, 80–83, 168, 169, 220
　　inevitability of, 185
　　in jobs, 185
　　making successful, 186
　　in medicine, 100, 120–24
　　multiple technologies as driving, 117
　　and nano world, 134–36
　　past as shelter from, 198
　　speed of, 6–8, 122, 134–36, 175–83, 185,
　　　223–24
　　and stability as mark of shame, 185
　　violence of, 6–8
　　welcoming, 225
　　See also genetic engineering
chess, 104
Chevron-Texaco, 41
Chicago, Illinois, 158
Chile, 150, 151, 152
China
　　cloning in, 211
　　and foreign students remaining in U.S.,
　　　203
　　and 427:1 gap, 19, 20–21
　　and genetic revolution, 101
　　language development in, 63
　　and migration of scientific literate people,
　　　191, 203, 208
　　natural resources in, 53
　　population in, 86
　　public health in, 86
　　as superpower, 224
　　wealth of, 53
chips
　　Affymetrix, 116
　　cost of, 35–37
　　early, 127
　　error tolerance of, 127
　　gene, 113, 116, 117–20, 217, 221
　　and nano world, 127

Rojitas (story), 195–96
romance languages, 64
Run (island), 55
Russia/Soviet Union
cloning in, 211
education in, 189
natural resources in, 48–49, 53
as superpower, 224
and wealth, 48–49, 53

S

Safra, José and Moisés, 42
salaries. *See* income
San Francisco/San Diego, California, 101,
158, 189
San Jose, California, 158, 191
Sanger Centre, 111
Saudi Arabia, 49, 53, 57
Schumpeter, Joseph, 172
science fiction, 11, 216
scientific literacy
and 427:1 gap, 24
and global economy, 28–29
importance of, 10
lack of, 11–12
and migration, 156, 191, 201, 203, 205–6,
208
and service economy, 29
and wealth/income, 25–45, 186–87
See also education; foreign students in
U.S.
Scopes, John, 190
Scott, Bruce, 50
Search for Extraterrestrial Intelligence
(SETI), 34
Sears Roebuck, 181
service economy, 28, 29, 30
Shell Oil, 57
Shelley, Percy Bysshe, 17
Singapore
cloning in, 211
education in, 51
exports of, 151
impact of change on, 7
knowledge creation in, 142, 152
knowledge economy in, 51
life-science initiative in, 101
and migration of workers, 152, 207

as superpower, 224
wealth in, 7, 27, 51, 52, 53, 57
"Singapore Law," 56
Slim, Carol, 42
Smalley, Richard, 128
SMASH (Simple, Many, and Self-Healing)
computer, 114
Smith, Ham, 222
Social Security, 193
software, 36, 114, 164, 202
South Africa, 49, 53, 189
South Asia, and 390:1 gap, 21
South Korea
economic growth in, 140, 188
education in, 187, 188
exports of, 151, 152
and foreign students remaining in U.S.,
203
knowledge creation in, 139, 140, 142, 151,
152
technology export in, 152
wages in, 27, 140
and wealth, 27
Soviet Union. *See* Russia
space, vacuum of, 132
Spain, 24, 59, 142, 207
species
genomes as similar across, 218–19
See also specific specie
stability, hemispheric, 156–57
standardization, of communication, 62–71
standardized tests, 189, 191, 194
Stanford University, 113
stem cells, 169
stock market, 141, 178
Sulston, John, 99, 108, 109–11
Sun Microsystems, 34, 201
supercomputers, 101, 104–5, 108, 109, 125,
129
Switzerland, 53, 142, 189

T

Taiwan
exports of, 151, 152
and foreign students remaining in
U.S., 203
impact of change on, 7
knowledge creation in, 142, 151, 152

[ALSO BY JUAN ENRIQUEZ]

Revealing
America's
Fault Lines

[Juan Enriquez's unique insights into the financial, political, and cultural issues we face will provoke shock and surprise—and lead you to ask the question no one has yet put on the table: Could "becoming untied" ever happen here?

What America will look like in fifty years depends on what we do today.]

The Untied States of America
ISBN-10: 0-307-23752-4
ISBN-13: 978-0-307-23752-1
$24.95 hardcover
($34.95 Canada)